U0192257

Architecture and Materials
建筑与材料

［芬］埃萨·皮罗宁（Esa Piironen）方海 著 | 张亚萍 译

中国电力出版社
CHINA ELECTRIC POWER PRESS

图书在版编目（CIP）数据

建筑与材料 / [芬]埃萨·皮罗宁（Esa Piironen）方海著；张亚萍译.
—北京：中国电力出版社，2021.9
ISBN 978-7-5198-3831-7

Ⅰ.①建…　Ⅱ.①埃…　②方…　③张…　Ⅲ.①建筑材料　Ⅳ.①TU5

中国版本图书馆CIP数据核字（2019）第250705号

出版发行：中国电力出版社
地　　址：北京市东城区北京站西街19号（邮政编码100005）
网　　址：http://www.cepp.sgcc.com.cn
责任编辑：王　倩（010-63412607）
责任校对：黄　蓓　郝军燕
装帧设计：锋尚设计
责任印制：杨晓东
印　　刷：北京博海升彩色印刷有限公司
版　　次：2021年9月第一版
印　　次：2021年9月北京第一次印刷
开　　本：710毫米×980毫米　16开本
印　　张：11.25
字　　数：185千字
定　　价：88.00元

版权专有　侵权必究
本书如有印装质量问题，我社营销中心负责退换

建筑中的材料灵魂

Soul of Materials in Architecture

Contents

Foreword   022

Introduction   024

Chapter 1   Warm as wood   036

Chapter 2   Bendable as bamboo   058

Chapter 3   Hard as stone   066

Chapter 4   Solid as brick   074

Chapter 5   Reinforced as concrete   090

Chapter 6   Thin as steel   106

Chapter 7   Transparent as glass   144

Chapter 8   Mouldable as plastics   156

Chapter 9   Foldable as fabric/textile   160

Postscript   164

Appendix   172

Index of Images   173

目录

序　形式追随材料　方海　　　　006

前言　　　　023

绪论　　　　025

第一章　温暖如木　　　　037

第二章　柔韧之竹　　　　059

第三章　坚硬之石　　　　067

第四章　方正之砖　　　　075

第五章　坚固之混凝土　　　　091

第六章　纤巧之钢　　　　107

第七章　通透之玻璃　　　　145

第八章　可塑之塑料　　　　157

第九章　可折之纤维 / 织物　　　　161

结语　　　　165

附录　　　　172

图片索引　　　　173

译后记　张亚萍　　　　178

序

形式追随材料
方海

人类近万年的文明史实际上就是人类与材料的交流与交互发展史。材料环绕着人类，材料让人类脚踏实地，材料与人类的合作发展出近乎无穷尽的物质文明。不同地区的不同民族在不同的历史发展阶段发现、发明并使用着不同的材料，从而发展出不同的文明。近东两河流域的美索不达米亚的先民们用太阳干焙下的砖版刻出人类最早的楔形文字；尼罗河流域的古埃及工匠则将其发明的象形文字刻在石版上，同时又用巨石雕出帝王形象并建造出石构建筑；而源于黄河和长江流域的中国夏商周先民们先后在甲骨、玉石、青铜、竹简和石板上发明了汉字，并发展出东方写实主义造型艺术。在地球的两端，分别发展出的立足于黄土高原的黄色文明和立足于地中海环岸古国的蓝色文明，引领着不同地区的人类文明相互独立地向前发展，经大航海时代全球范畴的文明交融，人类逐渐进入工业化、电气化、信息化的全球一体化时代，并由此开始提倡绿色文明。绿色文明呼唤绿色生态材料，建筑师、设计师和他们与材料的故事由此进入新的篇章。

1. 洪堡去寻找什么？

世界各地的古人当中有相当一批智者，在几千年前，面对旷野，仰望星空，已发展出朴素的关于人与大自然关系的指导思想，如中国古代哲人的"天人合一"论和古希腊哲学家的"自然论"等。然而，千百年来，因物质条件和科学发展程度的制约，人们很难确切而系统地全面考察和理解人与大自然的关系，因此，尽管欧洲各国在文艺复兴之后竞相在数学、天文学、物理学、化学、生物学、地理学和博物学方面都有长足的进步，却始终没有形成系统而生态的科学系统发展观。时代呼唤科学巨人的出现，于是，19 世纪世界最伟大的科学大师应运而生，他就是德国探险家和科学家亚历山大·冯·洪堡。

2015 年，著名科普作家安德里亚·沃尔夫（Andrea Wulf）隆重推出其新著《The Invention of Nature: Alexander Von Humboldt's New World》，该著至今依然是欧洲各国学术书店的畅销书。今天的中国出版业在这方面也能追随时代发展的步伐，后浪出版社和浙江人民出版社迅速推出边和的译本《创造自然：亚历山大·冯·洪堡的科学发现之旅》。在 18 世纪，人们对大自然还处于各自为政的观察、探索和改造当中，并在诸多得意的时刻逐渐失去对自然的敬畏，于是，"如何理解自然"成为各地正直而深刻的科学家们需要深入思考的问题。这个问题时刻呼唤着洪堡：他的内心激荡着不息的求知冲动，不仅渴望周游世界，更试图洞察整个宇宙，厘清人与大自然的系统关系；他坚持用双足丈量地球，同时也将科学与想象结合起来，用"生命之网"的整体视角重新审视大自然。

洪堡不仅被誉为 19 世纪最伟大的科学家，也曾被普鲁士国王腓特烈·威廉四世盛赞为"大洪水后真正伟大的人物"。他深入委内瑞拉的茂密雨林，穿越漫长的安第斯山脉，攀登当时公认为最高的火山——钦塔拉索山；他曾与同伴邦普兰惊险地逃脱鳄鱼之口，目睹野马与电鳗的残酷搏斗，在重重树影间与美洲豹狭路相逢；他将对大自然的崭新理解，融入对彼时政治局势的悉心体察，既为托马斯·杰斐逊带去详尽的考察资料，还影响了西蒙·玻利瓦尔的拉丁美洲革命；洪堡对世界的广博认识不仅震惊着拿破仑，更深度影响着歌德、柯勒律治、达尔文、梭罗、海克尔等诗人和科学家们。洪堡去寻找什么？他要为人类带来什么？他当然为我们呈现了关于大自然的深邃思考：地球是独一无二、环环相扣的有机体，人类的任何破坏都有可能导致灾难性后果。与此同时，洪堡更为欧洲的科学界和实业界带来了博物学的丰硕成果，以及材料科学大发展的重要保障。1804 年，当他在美洲完成了五年多的科学考察回到巴黎时，他带回了大约六万件的植物标本，包括 6000 余种植物种属，而其中有近 2000 种是欧洲博物学家从未见过的：这是非常惊人的数字，考虑到 18 世纪末的欧洲，人们已知的物种也只有 6000 余种。从此，欧洲园林的发展获得了更多博物学的可能性，欧洲的科学、技术与工程的发展更是突飞猛进，从植物学到生物学，从基本元素到无穷尽的合成材料，人类的物质文明从此如同长上翅膀，一发不可控制。

对建筑师和设计师而言，他们的每一步工作实际上都是与不同材料交集的故事，因此他们应该像科学家和工程师一样，对以材料科学为核心的各类科学和技术成就时刻挂念于心。材料的发展和物质元素的探索，从根本上决定着每一位建筑师和设计师及其构思的成败。2013 年英国科普作家约翰·布朗（John Browne）出版新著《Seven Elements that Have Changed the World》，该书三年后由中信出版集团出版，薛露然的译本《我们如何走到今天：改变世界的 7 种元素》。作者用 7 种元素演示 7 段历史，对地球上的元素及其演化而成的材料家族进行了发人深省的解读，从科学的视角解读元素，我们看到了质子、中子和电子的 7 种不同组合，赋予了 7 种元素完全不同的个性：代表着力量又具有进攻性的镍，组成生命又释放热量的碳，折射着人类贪婪欲望的金，为世界留下影像的银，蕴藏巨大毁灭能量的铀，创造白色纯净空间的钛，以及催生互联网世界从而彻底颠覆人类生活的硅。从元素到日用材料的演化史，是人类科技革新和工程进步的历史，同时也是人类破坏生态环境从而走向贪婪疯狂而无法自控的历史，元素与材料在促进人类繁荣的同时，也改变着人类的本性，引发人性中黑暗的一面。建筑师和设计师必须尽可能学习、了解并掌握材料性能，从而尽最大可能为不同材料和人类的真善美目标服务。

2. 中国建筑与欧洲建筑：形式追随材料

当一百多年前现代建筑开始酝酿成形时，芝加哥学派主将沙利文以"形式服从功能"作为现代主义建筑最基本的原则，当五十多年前后现代建筑逐渐生成时，作为主要旗手的后现代建筑大师文丘里则用"形式服从装饰"作为后现代建筑的基本特征；此后，高技派举起"形成服从技术"的大旗，解构主义宣传"形式追随表皮"的观念，而极简主义更是打出"形式源自冥想"的口号……在从现代建筑到当代建筑发展的不同阶段，各种理论和各种思潮都在左右着建筑师们的实践原则，形式与功能、与装饰、与技术、与表皮、与冥想的关系实际上都是一种辩证关系。在社会发展的不同阶段，形式可以服从功能，也可以服从装饰，可以追随技术，也可以追随表皮，当然也可以源自冥想。然而，当我们站得高一些，看得远一些，回望历史的长河，反思世界建筑的发展轨迹，我们会发现建筑最重要的发展规律其实是"形式追随材料"，在古代世界，不同的材料决定着不同地区的建筑形式和建筑发展规律。

"木构的篇章"和"石头的音乐"是人类的建筑进入现代社会之前人们对中国建筑和欧洲建筑的直观写照，这两大传统建筑系统是对"形式追随材料"的最好说明。至少三千年前，中国已发展出榫卯构架为主体的木构建筑体系，到秦始皇统一中国，以阿房宫为代表的中国古代建筑达到早熟，这种"木构架＋榫卯＋斗拱"构造体系，从此在中国根深蒂固，长达三千年无可撼动。每逢改朝换代，宫室烧毁重建，前赴后继的中华大家庭后代都基本继承秦汉时代成熟发展的汉民族木构建筑传统，北宋《营造法式》的颁布是中国古代木构结构系统成熟的标志，此后又有清代《工部营造则例》等官方建造文本，使木构建筑成为中国进入现代社会之前的主流建造模式。中国古代虽有砖石建筑，但大多都用于城墙和陵墓，远非中国古代建筑的主流模式。中国地大物博，建筑材料众多，为什么中华民族的祖先唯独选中木构作为中国古代建筑的基本结构形式？其缘由众说纷纭，但最有可能的说法则是中华民族的主要文化发源地黄河流域在两三千年以前一直是森林密布的地方，无边无际的原始森林加上中华民族早熟的木工工具发展，让我们的祖先很早发展出完善的木构模式，并在很早的时代就以国家制度规定建筑模式，从而使中国古代木构建筑在三千年的历史长河中以一种超稳定的方式存在，直到西方列强武力打开中国国门，强行输入现代建筑的思想和建设系统。

传统观念中的西方建筑源自古埃及和古西亚建筑传统，其主体都是石构建筑，保留至今的埃及金字塔和神庙时刻展示着西方石构建筑的巨大威力和永恒魅力。紧随其后的古希腊和古罗马文明则在全盘继承古埃及石构体系的基础上，又将石构建筑发展到一个崭新的高度，而古罗马更是发现并发明使用了混凝土，从而使西方建筑在材料的持久性和结构的稳定性方面居于领先地位。即使欧洲在古罗马帝国崩溃后进入中世纪的漫长岁月中，以罗马风和哥特风格为主流的欧洲中世纪建筑依然是黑暗中世纪漫漫长夜的一盏明灯，成为欧洲中世纪文明的最重要硕果之一，现存欧洲各地的每一座罗马风和哥特式教堂，以及城堡和庄园，都成为欧洲古代文明的瑰宝。随后源自意大利佛罗伦萨的文艺复兴运动，则是重新发现古希腊、古罗马古典建筑和艺术的规则和魅力，古典的石构建筑以更强大的气场出现在欧洲各国，此后的巴洛克风格、洛可可风格、新古典主义风格等，都是不同风格、不同装饰手法的石构建筑在欧洲各地的轮流表演。石构建筑构成西方建筑的核心要素，石材与木材截然不同的性能，决定着西方建筑与中国建筑之间完全两样的形式系统。东西方的园林也在"木构的篇章"和"石头的音乐"的对比中分别展现"太湖石与正面体"的独特魅力。太湖石的自然、随意与木构建筑的森林意向一脉相承，并由此引申为中国古典园林的自由式布局、小桥流水，以及用品种有限的花草植物编织园林主人的个性梦境。与此同时，正面体的规范、有秩则与石构建筑的几何抽象息息相关，并由此发展为西方古典园林中几何式布局、规整的园林建筑与水池，以及对博物品种的无限追求，来实现从园林看世界的宏大理想。中国与西方在建筑与园林方面的交流至少有五百年的历史，"木构的篇章"与"石头的音乐"时有交融，"太湖石"与"正面体"更是相互好奇与欣赏。于是英、法、德等欧洲列国在过去的四百年间不断出现"中国风"的建筑和园林，而号称"万园之园"的圆明园则是西方建筑与园林在中国的集大成者。然而，这种交流在工业革命之前都只能是猎奇之举，最终，西方现代建筑以压倒性优势征服中国，也征服全世界，而新型建筑材料则是这种全球征服的马前卒。

从根本上讲，现代建筑的起源、发展直到当今建筑的多元化面貌，其背后最基本的推动力，就是建筑材料及其相应结构系统的演化和发展。随着社会的发展和人口的激增，全国范围内的大城市应运而生，各种新型功能的建筑层出不穷，新型建筑需要新型结构，尤其是工业化革命之后大批工业建筑需要大跨度结构，而大企业和商业建筑的办公楼则需要高层结构。于是，传统的砖石结构和木构显然无法满足时代的需求，以钢、混凝土和玻璃为代表的新材料走向台前。水晶宫和埃菲尔铁塔的出现，使钢与玻璃展示出了强大的生命力，此后钢与玻璃成为现代建筑的主体材料。第一代建筑大师中，格罗皮乌斯和密斯都是研究与使用钢与玻璃的经典大师。与此同时，新型的钢筋混凝土则成为艺术表现力更强的现代建筑材料，第一代建筑大师中的赖特和柯布西耶以其惊人的创造力用钢筋混凝土完成了一系列现代建筑杰作。而芬兰建筑大师阿尔托从地域主义文化和生态观念入手，对芬兰和欧洲的传统木构建筑和砖石建筑进行研究梳理，创造出现代主义建筑的新流派，至今依然影响着全世界不同地区、不同民族的建筑师们。21世纪初德国出版的建筑设计手册中，其主题即是以材料分类的结构与构造系统，有《钢结构建筑设计手册》《混凝土结构建筑设计手册》《玻璃结构建筑设计手册》《木结构建筑设计手法》和《砖建筑设计手册》。此外，世界各地也分别出版有《石建筑设计手册》《铝建筑设计手册》《草泥建筑设计手册》和《竹藤建筑设计手册》等。新的时代随时呼唤新型的材料，这其中既有科技发展所带来的新型材料，也有对传统材料进行改良之后的材料运用。熟悉各种材料，精心研究材料，随时关注新型材料的进展情况，是现代建筑师的必修课。今天，我们面临着全球化的生态恶化趋势，提倡绿色设计势在必行。关注、研究和主动运用各种生态材料是当代建筑师和设计师的基本任务，而以中国为主导的合成竹材的开发和使用，正在吸引着国内外大量建筑师和设计师们，笔者在过去20多年一直潜心于合成竹材的研究及其在建筑、室内、家具和工业设计中的广泛应用，并与诸多芬兰建筑师和设计师展开合作，完成了一批以竹材研究与运用为导向的建筑与家具作品。

3. 中国竹材与当代建筑师

竹是中国最重要的天然材料之一，中国人已经与各类竹制器物至少共同生活了三千多年。中国各地不断出土的文物，尤其是从战国到秦汉的竹简，权威记载着中国传统文化的精华内容，同时也是中国书法早期的华丽篇章。从日常生活的各个方面来说，竹子对中国人如此重要：竹建筑、竹家具、竹器具、竹乐器、竹食品，以及用竹原料制作的中国宣纸。更引人注目的是，在竹简书法之外，墨竹也成为中国传统水墨艺术中最重要的门类之一，竹子由此具有独特的艺术与文化意义，并与梅花、菊花和兰花在中国画和其他艺术形式中并称为"四君子"，也与松树和梅花并称为"岁寒三友"。此外，竹子是中国诗词中无穷尽的永恒主题，其中最著名的就是北宋大文豪苏东坡的"宁可食无肉，不可居无竹"。

中国并非世界上唯一生长和栽种竹子的国家，却是最重要的发明并发展出完善的竹制建筑与竹制家具和日用器具的国家，其古老的设计智慧至今仍对现代建筑和设计诸多方面产生积极影响。千百年来，中国和世界各地的产竹国家都使用原竹，在简便、经济的同时，也难以回避防火、防蛀、防潮、防腐等问题。因此，从 20 世纪末开始，以绿色设计和生态科技为己任的中国科学家和企业家通力合作，开始了合成竹材的研究与生产。南京林业大学的张齐生院士和王厚立教授、杭州大庄竹业集团的林海先生等，都是中国当代合成竹材研究与生产的最重要代表。合成竹材最初主要用于室内地板及墙板，随后开始用于建筑室内隔断、家具、围护及声学材料，并从室内延伸至户外，现在则已进入合成竹材料和建筑结构的系统研发阶段。

笔者高中时代曾经临摹过一整本"元四家"之一吴镇的《竹谱》，至今对竹子的喜爱始终在延续与加强中。当竹子成为绿色设计时代非常引人注目的生态材料时，作为建筑师的笔者心中最强烈的愿望就是在自己的设计中最大限度地应用竹材。随着绿色设计理念在全球和在中国的不断深入人心，合成竹材的研发和使用发展的速度非常迅猛，笔者在 1999 年有幸参加国际竹藤组织委托的"传统竹产品的现代化设计及产业化研究"项目。在此期间与杭州大庄竹业有了愈来愈频繁的接触、交流与合作，笔者本人也成为我国第一位在建筑、室内和家具设计中全方位使用合成竹材的建筑师。2002 年笔者接受深圳家具协会邀请主持设计深圳家具研究院办公楼设计，并从一开始就决心完成中国第一座生态建筑。除了用生态设计的理念完成该设计的通风、采光和太阳能系统设计之外，该建筑最大特色就是从建筑、室内到家具领域大量使用经杭州大庄竹业多年研发的合成竹材。不同类型、不同规格的竹材，被用于建筑的内外墙体和避阳板，用于室内外地板和楼梯踏板，同时也用于家具等处。

深圳家具研究院建成于 2004 年，其合成竹材的使用立刻引起世界各地建筑师的广泛兴趣，对于时刻关注地球生态状况并大力提倡绿色设计理念的建造师来说，中国的合成竹材对全球的生态建筑都具有特殊的意义。此后不到十年间，世界各地就涌现出了一大批大规模应用杭州大庄竹业集团研发生产的合成竹材的著名建筑案例，其中最著名的包括英国建筑大师查理德·罗杰斯设计的西班牙马德里新国际机场。该机场用大庄合成竹材完成了 12 万平方米的地板和 16 万平方米的天花设计；美国建筑大师斯蒂文·霍尔设计的深圳大梅沙万科集团总部办公大楼，该建筑除主体结构是钢筋混凝土之外，其余内外装修和家具系统全部使用大庄合成竹材；芬兰建筑大师拜卡·萨米宁设计的无锡大剧院，该建筑在其主体剧场和音乐厅的室内全方位使用大庄合成竹构件，是当代单幢建筑使用合成竹材体积最大的建筑案例。

2018 年年初，芬兰 Parus Verus 出版公司由笔者和德国当代建筑评论家法克·雅格尔（Falk Jaeger）教授主笔撰写的《无锡大剧院》，全面介绍了芬兰建筑大师萨米宁教授和他的设计团队如何以中国文化传统为构思的起点，以材料研究为细节设计的出发点，最终将无锡大剧院打造成无锡城市新名片的故事。萨米宁是芬兰现代建筑第四代建筑师的杰出代表。从 19 世纪末开始的现代建筑和设计运动至今，芬兰建筑师在国际建筑和设计领域一直享有极高的声誉。芬兰第一代现代建筑师的代表人物伊利尔·沙里宁（Eliel Saarinen）以其前期的民族浪漫主义风格和后期的现代风格享誉世界，成为从芬兰到欧洲，再到美国，乃至全世界的设计导师。第二代芬兰建筑师代表人物阿尔瓦·阿尔托则因其成功地从现代主义国际式的设计教条出发开创出地域主义生态建筑风格并迅速影响全球，从而时常被誉为与赖特、格罗皮乌斯、勒·柯布西耶和密斯并列的"现代建筑五大师"之一。芬兰第三代建筑师的代表人物是一位非常独特的、非典型芬兰建筑师雷曼·比尔蒂拉（Reima Pietilä）。他深入细致的理论思考和与众不同的建筑思维手法，成功摆脱了老沙里宁和阿尔托的巨大光环。一方面关注当时的后现代建筑思潮，另一方面更注重全球化趋势下日益严重的生态环境问题，为之后芬兰和全世界建筑的多元化发展奠定了基础。萨米宁作为芬兰第四代建筑师的代表人物，面临着全球化、多元化、信息化的环境，从一开始就将对建筑材料的系统研究作为设计的基础课和出发点，同时也决心向美籍芬兰设计大师小沙里宁和丹麦建筑大师伍重学习，不再拘泥于对特定风格和设计手法的追求，而是从材料研究入手，从大自然和民族历史的相关元素中获取构思灵感，以此形成自己的创作模式。

萨米宁早在 1968 年就建立了自己的建筑事务所，他从创业伊始就不忘初心，坚定不移地开始了对材料的深入研究和对结构的广泛探讨，并且在其设计生涯的每个阶段都能获得业内的认可和国内外专业奖项。作为"刚从森林里走出的芬兰人"，萨米宁像大部分芬兰建筑师一样对木构和木材产生了浓厚的专注兴趣。他为自己设计的事务所办公室、工作室以及一系列住宅别墅，都获得了业内认可与好评。1970 年，他为芬兰著名设计师 Esko Pajamies 设计的都市别墅已被芬兰政府列为现代设计遗产。从 20 世纪 70 年代中期开始，萨米宁开始系统研究混凝土，并设计完成了一大批以混凝土为主体结构的学校、办公楼、剧院等公共建筑，其中拉赫蒂剧院不仅获得芬兰国家级混凝土建筑大奖，而且多次被相关专业杂志选为欧洲最佳混凝土建筑。20 世纪 80 年代中期的萨米宁进入其职业生涯的第一次高峰，其代表作则是赫尔辛基国际机场。也正是从这个项目开始，萨米尔正式开始了其对钢结构与玻璃构造系统的专业研究与探索，其结果是赫尔辛基国际机场不仅多次荣获芬兰国家建筑奖和钢结构建筑奖，而且连续多年赢得国际航空组织颁发的"世界最佳机场"称号。随后，在德国纽伦堡玛利亚音乐厅，萨米宁将其对钢与玻璃材料的研究推向高潮，在将经过第二次世界大战轰炸的中世纪大教堂改建为现代音乐厅的过程中，萨米宁用钢和玻璃谱写了一曲华丽的建筑篇章。建成后的玛利亚音乐厅最终获得更多奖项，包括芬兰国家钢结构建筑奖、德国国家建筑奖、欧洲最佳生态建筑奖等。进入新世纪之后，萨米宁追随着全球化的潮流来到中国，开始关注和研究中国的建筑材料和建筑模式，最终通过无锡大剧院走向其职业生涯的第二次高峰。

2015 年，芬兰国家建筑博物馆为萨米宁的新世纪中国建筑项目举办了第一次展览，这次展览的主题是"竹材与无锡大剧院"，整个展览除了无锡大剧院从构思草图到设计竞赛再到工程模型的整体展示外，其主体展品就是无锡大剧院建筑中大量使用的、各种规格各种类型的由大庄集团研发生产的合成竹材。作为无锡大剧院设计团队的建筑师之一，笔者除参与从设计竞赛到施工图合作的全过程之外，主要负责室内与家具设计中竹材设计环节。对合成竹材这种当代诸多建筑师充满兴趣的生态建筑材料从热爱到充满敬畏，使笔者最终与大庄竹业集团同仁一道完成无锡大剧院两个主体剧院规模庞大的竹制室内和天棚设计，在主创建筑师萨米宁的主持下，创造了竹材设计的辉煌。2018 年 4 月，芬兰国家建筑博物馆为萨米宁的中国建筑项目举办了第二次展览，这次展览的主题是"空间、材料与工艺"。除用简洁展板展示其在欧洲的部分代表性作品外，主要展品均为萨米宁事务所近年在中国的三个最重要项目的相关资料。这三个项目是无锡大剧院、成都云端大厦和福州的海峡两岸文化交流中心。整个展览延续 2015 年展览的设计风格，用大量不同尺度不同材质的模型和局部建筑构件，以及主体建筑材料样品充实各个展厅。海峡两岸文化交流中心是当代中国最大规模的建筑项目之一，包括大歌剧院、小歌剧院、音乐厅、文化剧场和地方剧场五个功能各自独立，但空间上又相互交接的文化建筑空间。萨米宁对建筑材料有天然的敏感和兴趣，尤其注重尽最大可能研究和使用本地材料，并将其视为建筑文化传承的重要组成部分。在福州项目中，萨米宁除了继续大量使用曾在无锡大剧院室内构造中担任主角的合成竹材之外，又特别研究了福建和中国南部沿海地区虽然盛产但却被中国建筑师长期漠视的陶瓷、玻璃和品类繁多的石材，最终用绿色生态设计的理念，将时尚的形式和图案融入中国传统的建筑材料中，使得海峡两岸文化交流中心的建筑本身就已成为现代建筑、材料展示、中国传统工艺与东西方物质文化交融展示的博物馆。

4. 中国竹材与芬兰设计师

"形式追随材料"的规则对设计师而言，与对建筑师相比，应该具有更深刻同时也更密切的意义。对设计师而言，设计的内涵和材料的情感，以及如何关注和处理两者的关系，既是任何设计的出发点，又是设计师的基本功所在。许多人们多年早已习以为常的材料，经过现代化的理解和加工处理，往往会焕发出全新的魅力，在这方面，中国竹材就是一个典型案例。笔者在过去二十多年的教学、研究和设计实践中，除参与前文提到的建筑项目外，完成更多的是合成竹材在家具和日用产品设计中的广泛应用。通过对合成竹材广泛、长期且深入的研发，探索和体会不同品类合成竹材的性能，并以此为出发点展开现代家具和日用品的设计。随后与芬兰建筑师和设计师长期合作，同时与中国工匠在制造工艺方面深度交流，最终研发并制作出"新中国主义"设计品牌下的现代中国竹家具和竹家居产品。

中国传统竹家具历史悠久，源远流长，时至今日，我们不仅在中国南方各省随处可见种类繁多、物美价廉的竹家具，即便在中国北方，竹家具也时常可见。中国的竹文化早已跨越最初的竹材生长范畴，向北方扩大种植领域，与此同时，竹家具受民众喜爱的设计模式甚至能导致诸多北方城市从南方购买原竹材料，而后在北方制作竹家具。以材料进行分类，中国传统家具可分为四大类，即漆家具、木家具（包括硬木和软木）、竹家具（包括藤制家具）和其他材料（包括金属、石头、陶瓷、布料等）家具。另一方面，在相当长的历史时期，竹器与玉器、漆器、青铜器和瓷器一道，在中国传统家居文化和物质文化中成为独具中国特色的器物设计文化典型。笔者在拙著《现代家具设计中的中国主义》中曾对我国传统竹家具中的竹座椅系列进行相关研究，观察到中国传统竹家具虽是全部使用原竹材料，但人们在长期生活和休闲中得以仔细观察并深入研究中国各地不同原竹的材料性能，精心使用原竹中不同部位，由此创造出大量充满创意和美感同时又全面满足人们日常功能的竹家具和日用器物，发展和完善了独具中国特色的功能主义设计，贡献"弯曲元素"和"堆叠结构"等极富现代感的设计原则。这些设计原则曾经激发了许多现代设计师的创作灵感，由此使一大批现代版藤蔓家具、钢家具和钢木家具应运而生。

然而，原竹如何才能够成为现代设计师得心应手的材料？如何解决原竹特性中的最基本问题：人类可以吃竹子，昆虫也能吃竹子？在过去的三十年间，以张齐生院士为代表的科学家和以林海先生为代表的企业家兼竹材专家一直在为解决这个问题而努力，然后借助于阿尔托早年发明木制胶合板的技术启发，开始在中国南方产竹地区生产竹胶合板和竹集成材，充分利用竹纤维的韧性和强度创造出可广泛用于建筑、景观、室内、家具和工程设计的一种新型绿色生态材料。这种新型竹材与原竹相比具有全新的现代特征：他们不仅防虫防火、防潮防腐，而且在多维向度都可以达到很高的强度和硬度，同时也有一定的弹性，可以根据设计需要处理成任何形状和尺寸。由于竹纤维在材料的单向度方面具有更大的强度和塑性，从而使合成竹材在硬度和受力强度上接近甚至超过许多种类的硬木，其弹性更是硬木不可比拟，从而使得合成竹材为现代家具提供了新的设计灵感。

中国传统家具三千年的发展历程为后世留下了非常宝贵的设计智慧，成为现代设计师取之不竭的灵感宝库。当中国漆家具和其他材料的家具为漆艺等中国传统工艺的应用开发不断创造精彩案例时，中国的木结构家具和竹藤家具则在功能主义的基础上开创出不同系列、自成一体的设计模式语言和构造方法系统，从设计原理、设计手法和设计案例诸方面为现代设计师提供了极具教益的参考。笔者自 20 年前开始研究中国古代家具系统以来，一直关注并系统研究中国古代家具中的图文资料和实物案例。随着合成竹材料研究的深入，我们强烈感受到合成竹材这种新兴生态材料的召唤，这种源自中国古老沃土的新型材料随即成为"新中国主义"设计理念的思考契机。它与中国古代的框架椅、折叠椅、摇椅、各种凳子、桌子、橱柜等实例融为一体，为我们打开了"新中国主义"设计品牌的大门。常言道：当局者迷，旁观者清。对于中国古代的设计智慧而言，西方设计师往往是最热心的"发现者"和设计灵感方面的受益者，"新中国主义"设计品牌的另一位主要开拓者就是芬兰当代设计大师库卡波罗。作为北欧人文功能主义设计学派的集大成者，库卡波罗以科学家的态度看待设计，以工程师的观念理解设计，以艺术家的情怀解读设计，同时更以医学界的审慎将人体工程学原理全面引入建筑、室内和家具设计中，最终形成我们进行"新中国主义"设计的最基本原则，即人体工程学、生态设计原理和美学原则。我们作为设计师对合成竹材的研究和使用，与我们的长期合作的工匠印洪强先生及其工艺团队的工作密不可分。我们一同考察大庄竹业集团，从原竹切割到单板组合再到热压合成各类合成竹材的全过程，精确选择适合"新中国主义"设计的竹材，最终研发并生产出一大批基于竹材性能和中国传统设计原理的新型中国现代家具。在笔者最近出版的另一本新著《新中国主义设计科学》中，笔者系统介绍了十三个系列的"新中国主义"现代家具产品，即龙椅系列、龙椅茶几、图案框架椅系列、框架椅系列、小靠背椅系列、新中国主义竹靠背椅、合成竹靠背椅系列、儿童椅、吧台椅系列、多功能凳系列、阅读椅系列、主题装饰休闲椅系列和软包休闲椅等。合成竹材为设计师提供的设计可能性是无限的，可以让设计师尽情地展开想象的翅膀。

2005 年，笔者与芬兰著名建筑师和设计师威沙·洪科宁一道，受成都高新区管委会邀请设计成都高新区天府国际社区教堂。从设计的初始，合成竹材就在我们心目中占有非常显著的地位，因为我们已在深圳家具研究院的建筑、室内和家具设计中全方位使用合成竹材，同时也在无锡大剧院的设计中考虑大尺度合成竹材的应用，因此我们非常自然地希望在成都天府国际社区教堂中再次使用合成竹材。除了教堂办公区域内的常规家具继续沿用以前研发生产的"新中国主义"品牌设计之外，我们对教堂主体礼拜厅内的忏悔桌、忏悔跪凳及教堂长椅进行了专门设计。在忏悔桌的设计中我们将合成竹材的强度、可塑性和竹纹肌理诸性能做了极大的发挥，创造出现代雕塑感极强的宗教家具，再辅以色彩的涂饰，又将宗教的故事和含义酣畅淋漓地注入合成竹板制成的忏悔桌中。人们可以通过忏悔桌中不同色彩、不同形态的孔洞，不同程度地感受天国的幻影和人生的多变。在教堂长椅的设计中，我们试图从光与黑暗的辩证关系中寻找长椅的设计规则，最终将该系列的长椅设计成一系列独特且同时具有灯具功能的教堂座椅，再次充分利用合成竹材的优异性能制作出一系列与众不同的座椅。在大部分宗教仪式中，人们需要间接光源，有时更需要隐秘光源，在这方面，我们设计的教堂座椅为人们带来对室内光源和亮度的不同理解和使用的可能性：在欢快的婚礼会场，主礼拜堂的天棚与墙体上的灯光全部打开，座椅上的竹质光泽细腻宜人；在某些特殊的宣教仪式中，人们渴望黑暗的基调，关闭天棚和墙壁照明灯，此时，竹制座椅上安置的间接光源为整个空间带来了温馨的气氛和宗教的气场。

5. 艾萨·皮罗宁的材料情结

笔者认识芬兰著名建筑师艾萨·皮罗宁教授已有二十多年了，大约15年前笔者主编出版了皮罗宁建筑作品集，现在则正主编出版六卷本的皮罗宁建筑全集以及他的新著《建筑与材料》。这些天，当笔者着力思考建筑与材料和形式的关系时，刚好看到北京联合出版公司的科普著作《迷人的材料》，于是很快回想起与该原著（英文版）相遇的情景。2013年夏天快要结束的时候，笔者又与皮罗宁约好在位于赫辛基市中心学术书店二楼的阿尔托咖啡厅会面聊天。那天皮罗宁教授手里拿着一本书走进来，于是我们当天的聊天便从这本关于材料科学的科普书开始广泛谈及建筑师、建筑、材料和形式的关系，并由此启发皮罗宁教授决心记下自己半个多世纪作为建筑师与各种材料的情缘，最终成为《建筑与材料》这本小书的缘由。

皮罗宁教授那天带来的书就是英国当代著名科学家马克·朱奥多尼克（Mark Miodownik）于2013年出版的《Stuff Matters:The Strange Stories of the Marvellous Materials that Shape Our Man-Made World》。作者凭借自己对材料科学的精湛知识，信手拈来地从身边众多事物中取出10种材料进行深入浅出的讲解，它们是Steel，Paper，Concrete，Chocllate，Foam，Plastic，Glass，Graphite，Porcelain

和 Lmplant。五年后，北京联合出版公司出版了赖盈满译本《迷人的材料：10 种改变世界的神奇物质和它们背后的科学故事》。该书的序章一下子就能把读者带入神奇的材料世界，并让读者立刻明白一些看似简单，实则对建筑师、设计师和普通民众都非常重要的道理，如"材料构筑了我们的世界""文明时代就是材料时代""看不见的微观世界影响大"等。中文版的目录非常引人入胜：不屈不挠的钢，值得信赖的纸，作为基础的混凝土，美味的巧克力，不可思议的发泡材料，充满创造力的塑料，透明的玻璃，坚不可摧的石墨，精致的瓷器，长生不死的植入物质等，使读者迅速体会材料科学的魅力和趣味。作者在后记中，从五个方面总结材料科学之美，即"万物都由原子组成，结构尺度影响大，肉眼可见的尺度，生命与无生命的分野，以及材料拥有意义"。皮罗宁教授认为该书值得建筑师和设计师认真阅读，以便从更深层面理解材料，从而更深刻理解建筑。

皮罗宁同萨米宁一样，同属于芬兰现代建筑伟大传统的第四代传人，不过，与萨米宁相比，他属于阵容强大的芬兰第四代建筑师群体中相对年轻的新生代力量。第一代芬兰建筑师最显著的特点之一就是对材料的极大兴趣。与萨米宁在设计生涯的每一个阶段主要以某一种材料作为探索和应用主体的实践方式不同，皮罗宁热衷于在自己设计生涯的各个时期都尝试使用不同材料，这也许与其不同的专业教育背景有关。在芬兰，由于其强势而优越的教育体制和世界公认的建筑与设计方面的教育声望，绝大多数芬兰建筑师如第一代的沙里宁、第二代的阿尔托、第三代的比尔蒂拉和第四代的萨米宁等，都毕业于芬兰建筑学院，由此形成芬兰教育的优势并代代相传，至今依然在全球保持领先的地位。皮罗宁的教育背景使他在某种程度上可以被称之为非典型芬兰建筑师，因为他在芬兰建筑学本科毕业后，去美国完成了三年的硕士学位的学习，并对美国建筑系统考察，而后又回到芬兰，一方面进行职业建筑师资格学习和研究，另一方面开始设计实践，并逐渐成为后来第四代芬兰建筑师的优秀代表之一。

虽然皮罗宁的专业教育背景并非典型的芬兰式，但他进入设计实践后不断形成的建筑风格却是芬兰当代建筑风格的典型范式：人文功能主义理念下的精致而严谨，对材料系统研究下的规范而细腻，以及现代艺术创意情景下的多彩而浪漫。从最初以交通建筑和公共建筑为主体的建筑类型实践开始，皮罗宁即开始了对多种不同建筑材料的浓厚兴趣，首先是由芬兰传统木构建筑引发的对现代木构建筑的长期兴趣，其次是同样被认为是芬兰民间建筑传统的砖混建筑传统，以及随后出现但很快成为皮罗宁设计特色的钢结构。皮罗宁早期作品中一方面倾听业主对材料的意见，另一方面也积极为业主建议他喜爱的材料和结构。然而，皮罗宁中期的作品则更多地专注于混凝土结构，这一方面是因为国际上柯布西耶使用混凝土的浪漫气质所产生的全球化影响，另一方面更是因为芬兰自身所拥有的几位世界级混凝土结构大师，如阿尔瓦·阿尔托、布隆姆斯达特（Aulis Blomstedt）、比尔蒂拉和路苏沃里（Aarno Ruusuvuori）的作品对芬兰和世界各地的青年一代建筑师都产生了广泛影响。与此同时，随着皮罗宁专业上的成熟，他开始在芬兰频繁而公正的建筑设计竞赛中不断赢得大型公共建筑项目，而混凝土结构恰恰是钢与玻璃结构技术成熟以前最适合于大型公共建筑的材料和结构。在获得芬兰国家建筑奖的坦佩雷剧院项目中，皮罗宁对混凝土材料进行了透彻的研究和运用。但在该歌剧院的辅助建筑部分，皮罗宁开始钢结构的大跨度使用，并从此开始刻苦钻研钢结构，从而成为当代芬兰最重要的钢结构建筑专家之一。皮罗宁最重要的钢结构建筑代表作是赫尔辛基中央火车站的屋顶增建项目。赫尔辛基中央火车站是世界级建筑大师老沙里宁的成名作，也是全世界最著名的火车站建筑，虽然早已列入芬兰国家级文物保护单位，但一百年来始终是芬兰最繁忙的建筑。20世纪90年代后期，芬兰铁路及城市交通的高速发展迫使政府不得不作出决定：增建该火车站大天棚以便扩大其客容量，同时也改善其环境条件。但作为芬兰国宝的这座中央火车站增建项目，如何才能获得政府的首肯和民众的拥护？这种情况下，一贯以建筑设计自信的芬兰第一次决定举办欧洲范围内的建筑设计竞赛，除芬兰和北欧建筑师均可参赛外，还特别邀请著名的英国建筑大师福斯特、西班牙建筑大师莫耐欧等加入竞赛，可谓高手如林。最终结果，皮罗宁以其对钢和玻璃极其系统的研究赢得了这项重要的国家竞赛。新设计不仅需要结构与外观本身的美观和坚固，轻巧并理性，而且还要与老沙里宁的经典建筑作品有深刻的情感呼应和逻辑交融。此外，建筑师还必须设计出完美的施工方案，能够保证整个施工过程对建筑物本身和对火车站的日常运作基本没有影响。皮罗宁的设计圆满完成了这座经典火车站的增建，满足了所有的功能需求，受到国内外几乎所有使用者的赞扬，皮罗宁对钢和玻璃的驾驭在该项目中表达得炉火纯青，此后，整个芬兰开始进入钢和玻璃的黄金时代。

皮罗宁是一位非常敏锐同时也非常勤奋的建筑大师，从青年时代的求学生涯到成年后的设计实践时代，他无数次穿行于欧、美、日多国之间，有缘拜会现当代一大批建筑界著名人物并随时记录其交流内容，最后集结为《On Architecture》在芬兰出版，后又与笔者合作，在中国出版《建筑哲思录》。近几年，毕罗宁教授在广东工业大学和国内其他高校教学之余，开始将其对建筑与材料的理解记录下来，其记录一方面反思自己近六十年设计实践中对建筑材料的理解、探索和运用，另一方面也广泛思考建筑史中不同建筑师对材料精彩纷呈的思考和运用实践，相信对建筑师、设计师和相关领域研究者能有新鲜的启发。

Foreword

We see daily a lot of building materials in our environment, at least if we live in a city. Sometimes there is a visual chaos. We also touch many materials daily.

Materials belong to our life and buildings. Architects choose materials designing buildings. Do we know enough materials?

How the materials have an effect on human life? Are materials healing our environment or are we building more sick buildings? That means also more sick people.

We have to remember that architects are designing for human beings.

This book tries to give some more information on Materials in Architecture.

The text and images are mostly based on my lectures in China during the teaching years.

Esa Piironen

生活在城市中，我们每天都能在周围环境中看到很多种建筑材料，有时候甚至材料多到让人产生视觉上的混乱感。

我们每天也在接触各种建筑材料，建筑材料属于建筑，并渗透到我们的生活中。

建筑师设计建筑选择材料时，是否具备足够的建筑材料知识储备？

材料如何影响人类生活？建筑材料能治愈我们的环境，或是我们会建造出更多不健康的建筑？这意味着将危害更多人的身体健康。

建筑师们必须牢记自己是为人类设计建筑。

本书试图详细论述各种建筑材料。

书中的文字与图片来自我在中国教学期间的讲座。

埃萨·皮罗宁

Introduction

All materials have their own unwritten
laws.

Tapio Wirkkala

Our planet was born out of the Big Bang less than 4.5 billion years ago. The universe (cosmos) was born a little under fourteen billion years ago. alongside matter and energy. Since the Big Bang, all natural materials have continued their lives, transforming over time.

The material is an accumulation of particles which all follow their own rules of physics. Architecture cannot exist without the material. The relationship between an idea and the material is an object of ongoing investigation. Volumes have been written about materials. Much of this literature focuses on the physical qualities of materials. This knowledge is of great importance to builders and designers. There is, however, less research on the impact of materials on people's wellbeing. The general principle goes that the right materials should be used in the right place in a building. The way materials interact depends on the qualities of these materials. In architecture, however, the idea is more important than the material.

绪论

> 所有材料都有自己的特
> 性与法则。
>
> ——塔皮奥·维卡拉[1]

我们的星球诞生于 45 亿年前的宇宙大爆炸，而宇宙诞生于 140 亿年前。从宇宙大爆炸开始，所有的自然材料就一直在随着时间的推移持续地生长与进化。

材料是一种有着自身物理学特性的原子的聚集物。没有材料，就不存在建筑。设计想法与设计材料之间的关系是一个持续的互为研究对象的关系。学术文献中关于材料的部分，多数集中在研究材料的物理特性。这些知识对于建造师和建筑师无疑非常重要。但是，鲜有文献研究材料对人类健康的影响。通常的设计规则是：适当的材料运用到建筑的适当位置上。建筑材料间的相互影响取决于建筑材料的品质。然而，在建筑学领域，建筑设计的想法比建筑材料更重要。而且，真正高质量的建筑的产生是建筑材料与建筑设计想法的完美综合。

1　塔皮奥·维卡拉（Tapio Wirkkala，1915–1985）出生于芬兰汉科（离设计师重镇赫尔辛基 130 公里处，一个通行双语的小镇），或许环境因素使然，他在商业作品设计上以"多变"的天赋著名。塔皮奥·维卡拉擅长的设计类别琳琅满目，从玻璃制品、钞票甚至连绘画图像。在他的职业生涯里，参加过许多国际性的重要展览，并且担任赫尔辛基美术与设计大学的艺术指导多年，同时也赢得许多设计大奖，包括 1951 年及 1954年米兰三年展的三项金牌奖。"一生，至少要创造一个经典"，塔皮奥·维卡拉以奇想天外的概念，设计出令人赞叹的"水珠晶球"系列。他毕生提倡的"简约中求变，变化中极简"原则一直深深地影响着北欧及全世界的设计师。（译者注）

And yet it is the synthesis of the material and the idea that generates truly high–quality architecture.

In the early days of the common era, the basic materials used in architecture were earth, air, water, fire and ether. In the early days of construction, materials were processed by hand, using simple tools. Surfaces were often left quite robust; for example, honing stones smooth was hard work and unnecessary from the practical perspective. Sculptors would treat their surfaces to a smooth finish.

The earliest dwellings were built from wood, branches and leaves. Caves also served as early forms of shelter. In that period, only natural materials were available for building.

Now, alongside wood and stone, the most common building materials include concrete, brick (ceramic tiles), steel (metals), glass and plastic. A host of synthetic materials have been introduced to complement natural materials, and we are only now learning how to use them.

Global warming brings its own challenges to the use of building materials around the world.

Today, architectural materials are divided into, for example, natural and synthetic, or man–made materials. Natural materials still remain the same: stone, wood (paper) and water (ice).

Architectural materials development by humans include metals (steel, aluminium, copper, bronze, titanium, cast iron), ceramic materials (brick, ceramic tiles, glass), polymers (plastics, rubber, elastomer), concrete, textiles (cotton, nylon, wool), composite materials (reinforced plastic, carbon fibre, glass fibre).

The purpose of architecture is to bring the world of material into harmony with human life.

Alvar Aalto

人类社会早期的建筑主要使用的材料有泥土、空气、水、火等自然材料。

早期的建筑材料与建筑结构都是利用简单的工具用手工制作。建筑表皮一般都很粗糙：将石材表面磨光滑，从实践的观点来看是非常艰难和没有必要的。只有雕塑师才需要将材料表面处理到光滑为止。

人类最早的住所使用木头、树枝和树叶建造。洞穴是人类最早的庇护所。在那一时期，只有自然材料可以用于建造。

现在，除了木材和石材，最普遍的建筑材料还包括混凝土、砖、金属材料、玻璃和塑料等。作为自然材料的完善与补充，大量的合成材料已被引入运用到建筑上，我们如今正在学习如何使用这些人工合成材料。

全球气候变暖给全世界在合理运用建筑材料方面带来了的巨大挑战。

当下，建筑材料可被分为自然材料和人工合成材料（或人造材料）两大类。自然材料依旧是石材、木材（纸）、水（冰）。

人类开发的人工建筑材料有金属材料（钢、铝、青铜、紫铜、钛、铸铁等），陶瓷材料（砖、瓷砖、玻璃等），高分子材料（塑料、橡胶、合成橡胶等），混凝土材料，纤维材料（棉织物、尼龙、毛织物等），复合材料（强化塑料、碳纤维、玻璃纤维等）。

建筑的目的是使材料世界与人的生活相和谐。
——阿尔瓦·阿尔托

Architects study through their work the nature and qualities of materials in different conditions, and how different materials would interact. The measurable qualities of materials, such as the weight, hardness and durability, are determined by the laws of physics. The characteristics of a material are also subject to sensory evaluation, and the ageing, patina, sheen and other similar factors have an impact on the value of a material. The valuation of materials has varied quite substantially over time.

The feasibility of a material in construction depends on its strength, hardness, viscosity, impact resistance, density, corrosion resistance, heat and electricity conductivity and many other such qualities in addition to availability and price.

The use of materials is also governed by the ideal of using materials sparingly. Waste in material use is usually a sign of bad design.

The recycling of materials has also come back into the equation with the rise of ecological thinking.

For designers, personal familiarity with a material is important. It is important that designers have hands−on experience of different materials, not merely information obtained from literature. Experience in materials can only be accumulated by using them, through trial and error.

Not all materials can stand the test of time. Some materials age quicker than others. This should be taken into account in the design process as well as budgeting (initial costs vs. running costs). The surface treatments also affect the ageing of a material significantly.

Architecture is a multisensory experience, felt in the whole body. This adds another dimension to different materials and the way they can be used. The thumping sound of rock, the clanking sound of iron, the clinking sound of glass and the cracking sound of breaking plastic. In the end, it is the appearance that largely determines the use of materials, that is, the personal preferences of the designer and the client.

建筑师通过他们的工作研究不同条件下各种建筑材料的本质和特性，以及不同的材料之间如何相互影响。材料可以计量的特性如重量、硬度、耐久性是由材料的物理特性决定的。材料的特征还会受到感官体会的影响，如材料的老化程度、光泽、光滑程度以及其他相似的因素是衡量材料特征的重要指标。

材料在结构中的使用除了考虑材料的普及度和价格外，还取决于材料的强度、硬度、黏性、抗冲击度、密度、耐腐蚀性、导热导电性等这一类的品质。

材料使用必须遵循集约化、充分利用的原则，材料的浪费使用是设计失败的一个信号。

随着环保思潮的崛起，可循环材料再次回归。

对于设计师而言，熟悉材料非常重要。设计师们必须有对不同材料的实际设计经验和使用经验，而不仅仅是从文献上获得的理论信息。材料运用的经验积累只能通过使用材料来实现，反复试验各种材料，从失败中找到解决办法。

不是所有的材料都经得起时间的考验。有一些材料老化速度过快。这些必须在设计过程中和成本预算中（包括初始成本与运行成本）考虑到。表面的处理同样会对材料的老化产生重要影响。

建筑是一个多重感官的体验对象，需用全身来感受。这增加了不同材料及其不同使用方式的另一个维度。石材的砰砰敲击声、铁的咣啷碰撞声、玻璃的叮当声、塑料开裂的咔嚓声。但最终还是材料的外观在很大程度上决定了材料的使用，即设计师和客户的喜好决定了材料的使用。

The intrinsic meanings held by materials are reinforced by their impact on the acoustics of a space.

Our childhood experiences have a strong influence on how we feel about materials as adults. The way we feel about materials is associated with the seasons, more or less successful experiences with different materials, how materials age, where we come across them, their availability and their use around the world, traditions, and the ease of use.

It is important to recognise that we experience materials through all our senses: the sense of vision, hearing, touch, smell and taste all process information, including materials, in their own way. Furthermore, the way our senses interact also affects the way we perceive materials. There may even be a sixth sense, or even more.

Machines introduced by the industrial age also revolutionized the use of materials in construction. With machines, the surfaces could be more easily finished. New materials, such as steel and glass, gained significant ground in construction and architecture.

Today, we are looking to the future of construction. Development work around new materials is prolific. Sustainable materials are introduced to the market on a continuous basis. Trials also produce non–viable materials.

Building is becoming increasingly automated. In some cases, robots carry out some of the building work, just like in the automotive industry. The need for human input in building work is decreasing. Information technology helps achieve more accurate and reliable results in architecture, too.

Buildings are built from different materials. An architect has to choose most suitable materials to the buildings he is designing; the best material to a certain point of a building. A building consists normally the roof, the walls, the floor and the base or basement. For all these parts, there are lot of possibilities to choose materials. Depending where an earth you are working, materials to choose for these parts of a building differ a lot.

材料对空间音效的影响增强了材料的内在意义。

我们儿童时期的经历会强烈影响我们成年后对材料的感知。我们对材料的感知与季节有关，对于不同的材料，我们或多或少都有一些成功的体验，包括材料的年代，我们遇到材料的地方，它们的可用性和它们在世界各地的使用，材料的传统和材料的舒适性。

认识到我们是通过所有感官来体验建筑材料，这一点很重要：对于建筑材料，视觉、听觉、触觉、嗅觉和味觉都以自己的方式处理信息。与此同时，我们感官的相互作用也影响着我们感知材料的方式，甚至可能有第六感，以及更多。

工业时代机器的使用也推动了建筑结构中材料使用的革新。使用机器后，建筑表皮得以更简易地进行表面处理。金属、玻璃等新材料占据了建筑设计和结构设计的重要地位。

今天，我们正在探索未来的建筑结构。新材料的开发，成果丰富。可持续材料被不断地引入市场。同时也试验出了很多不适合使用的材料。

建筑行业正日益自动化，有时，机器人可以完成一部分建造工作，如同汽车制造工业一样。建筑行业中对人工的需求正在缩减，信息技术的发展同样有助于建筑行业取得更为精确和可靠的建造成果。

建筑由不同的材料建造而成，建筑师在设计建筑时必须选择最合适的建筑材料；在某种程度上最合适的材料匹配给最合适的建筑。建筑一般包括屋顶、墙面、地面和基础。所有这些部分，每个部分在选择材料上都有许多种可能。在不同的地区设计建筑，在建筑选材上也会有很大的不同。

Exterior materials of a building must last as long as possible in a normal use, but weathers are very different in various parts of the world. Temporary buildings have their own rules.

Also the interior of a building must have habitable materials; healing materials if possible. An architect has to know how materials behave in our environment and how the materials can be put together as well.

Designing details is very important part of designing buildings. The joints in a building are many times the weakest parts of structures. Architects should also use so called real materials and not artificial substitutes. The combination of a real material and a substitute is a crime to some critics.

And when using a real material, use it as naturally as possible based on materials own laws.

An Italian architect Carlo Scarpa adviced his students to use most valuable materials only minimally and cheaper materials more.

Throughout history, human beings have strived to give meaning to their existence by building structures that push the limits of their technical abilities. The role of different building materials has evolved strongly in the course of history.

Architecture has moved on from the time when form was emphasised at the expense of materials. Analysing the meaning of materials has again become topical.

Architect Louis Kahn once asked a brick what would it like to become; the brick wanted to be part of a vaulted arch.

Buckminster Fuller, in turn, asked how much did a building weigh; he favoured light–weight buildings. The Australian architect Glenn Murcutt channels the legacy he has adopted from Aboriginal culture: tread lightly on this planet.

世界不同地区的气候差异非常大，建筑外部材料必须在正常使用的前提下保证有足够长的使用寿命。而临时建筑则另当别论。

建筑内部材料则必须使用宜居材料，尽可能采用健康治愈性的材料。建筑师必须了解材料在环境中的使用规律以及不同材料间的整合使用。

细节设计是建筑设计的重要部分。很多时候建筑节点是建筑结构中最脆弱的一环。建筑师应使用所谓的自然材料而非人造材料，对于一些评论家而言，将人造材料和自然材料结合在一起使用就是犯罪。

使用自然材料时，要根据其自然材料的特性尽可能合理地使用。

意大利建筑师卡罗·斯卡帕（Carlo Scarpa）建议他的学生们尽量少使用昂贵的材料而更多地使用廉价材料。

纵观历史，人类通过不断突破技术能力极限来建造建筑，从而力求谱写下人类存在意义的篇章。在这一历史过程中，不同建筑材料的作用也发生了强烈的演变。

曾几何时，建筑开始忽视材料而强调形式，于是，建筑开始得到异样的发展。如今分析材料的含义则再一次变为热门话题。

建筑师路易斯·康（Louis Kahn）曾探寻一块砖想成为什么——砖能成为拱顶的一部分。

接着发明家巴克敏斯特·富勒（Buckminster Fuller）开始探索如何计算一座建筑的重量，他更喜爱轻巧的建筑。澳大利亚建筑师格伦·默科特（Glenn Murcutt）设计时融合继承了澳洲土著文化传统：轻行于这个星球上。

The biggest challenge in building is to defy gravity. Architectural expression is based ultimately how to resolve the matter of bearing gravitational forces through structures to the foundations and the ground.

Climate change and an awareness of the limits of environmental capacity have recently raised new perspectives concerning building materials.

Architecture is a synthesis of the material and the spirit.[1]

The three art forms of architecture, painting and sculpture are linked to one another in that they are all manifestations of the human spirit based on materia.

Alvar Aalto

建造最大的挑战来自于对抗重力，建筑的表现形式最终建立在如何通过建筑基础结构和地面结构解决建筑承重的基础上。

气候变化和环境承受极限意识的觉醒又掀起了一些关于建筑材料的新观点。

建筑是物质材料和精神涵义的综合体。[1]

建筑、绘画、雕塑这三大艺术形式相互关联，都是在物质材料基础上人类精神的表现。

——阿尔瓦·阿尔托

Chapter 1　Warm as wood

The job of the architect is to find the right material for the right place.

Renzo Piano

Wood is an ecological and recyclable building material. It is a renewable natural resource that continues its life cycle through genes, just like human beings.

Wood needs to be used as fairly bulky building elements in order for it to bind atmospheric carbon dioxide. However, the most efficient way to bind carbon dioxide from the atmosphere is through forests throughout the world, by leaving them standing. This is where experts may disagree.

According to some estimates, forests cover one third of the earth's land area. Some 45,000 different tree species grow in these forests, the majority of them deciduous.

Wood is used for many purposes. It is one of the most important building materials in many parts of the world, as it can be easily obtained, it is lightweight, durable and easy to work. Wood–built houses are common especially in the Northern hemisphere.

The heat insulation capacity of wood is superior to that of concrete and especially steel.

第一章　温暖如木

建筑师的工作是为合适的
场所发现合适的材料。

——伦佐·皮亚诺

木材是一种生态可循环的建筑材料。它是一种可再生的自然资源，如同人类一样，通过基因遗传实现生命传承。

大量的木材被用作建筑原料来吸收空气中的二氧化碳。而吸收空气中二氧化碳最有效的方式是全世界各地未被砍伐的森林树木。这就是专家们对是否使用木材的争论所在。

据估计，森林曾覆盖了地球三分之一的面积。大约 45000 种不同种类的树木生活在这些森林中，这些树木大部分是落叶树。

木材用途广泛。由于木材较易获得、重量适宜、耐久及易加工，它是世界上大部分地区最重要的建筑材料之一。木建筑，尤其在北半球非常普遍（图 1-1）。

木材的保温性优于混凝土，尤其优于金属材料。

Moisture causes wood to swell up, and continuous moisture encourages mold and rot to grow. Wood will begin to decay in humidity that remains above 20% for a long period of time.

It takes a couple of months for mold to start growing in wood if the relative humidity around it stays above 80% during this time. For the molding and decaying of wood to occur, the temperature must remain between 0 and 40 degrees Celsius. Wooden structures used in wet spaces can be protected with an environmentally friendly heat treatment or chemical pressure impregnation.

According to recent studies, the use of wood in interiors has a surprisingly positive effect on people. The research says that wood is a material which underpins good health and supports recovery.

Using wood can influence mood and stress levels. The studies show that people react favourably to wood, both physiologically and psychologically. Wood finishes make rooms feel warmer and more cosy, and induce a feeling of calmness.

Wood seems to have the ability to regulate the body's stress levels. In a comparison of different work spaces, stress levels measured as the electrical conductivity of the skin were lowest in rooms with wooden furniture. No corresponding calming effect was observed in rooms with white furniture and indoor plants.

Touching wooden surfaces gives a soft, safe, natural feeling. On the other hand, touching stainless steel, cold plastic, or aluminum at room temperature causes an increase in blood pressure. The studies did not observe a similar reaction when touching wooden surfaces. The research indicates that the positive effects cannot be achieved by using imitation wood.

水分会导致木材膨胀，持续的高湿度会使木材发霉和腐烂，空气中长时间保持 20% 以上的湿度，木材就会开始腐烂。

木材周边的相对湿度一直保持在 80% 以上，数月后木材会开始长霉。木材出现霉变和腐烂的温度范围为 0～40℃。适应了潮湿环境的木结构有利于对抗热力和化学侵蚀。

据近期的研究表明：室内环境中木材的使用对人们有惊人的益处——室内环境中的木材有助于人类健康和病人康复。

木材的使用影响着人们的心情和压力水平。研究表明，人们从生理上到心理上都表现出对木材的喜爱。木质面材使房间感觉更温暖与舒适，让人感到平静。

木材还有缓解人们压力的作用（图 1-2、图 1-3）。通过皮肤导电率测算压力水平，不同工作空间的比较研究表明：木质家具空间中人体的压力水平最低。在白色家具和室内植物构成的空间中没有发现相应的安抚效应。

触摸木质表面给人柔软、安全和自然的感受。研究发现，室温条件下触摸不锈钢、低温塑料或者铝材则会导致身体血压上升，而触摸木质表面时则没有发现相似反应。该研究表明，使用仿木材质无法获得与木材质同样的正面效应。

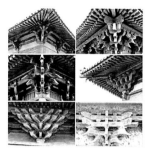

图 1-1　Old Chinese wood architecture 中国传统建筑木结构

图 1-2　Petäjävesi Church, Jaakko Leppänen, 1764　佩塔亚维西教堂，詹姆斯·列帕宁，1764

图 1-3　Shukagui Teahouse, Japan, 1659　虞姬茶室，日本，1659

Physiological measurements show that quality of sleep and recovery from stressful situations were better in a room finished in wood than in a room finished in imitation wood.

Positive psychological effects have also been observed in schools. Stress peaks measured by pulse variations in classrooms finished in solid wood die away soon after pupils arrive at school, whereas in control classrooms, moderately stressed conditions continue all day long. Pupils' experiences of stress and feelings of tiredness and lack of achievement were less in wood–finished classrooms than in normal classrooms.

The effect of using wood in interiors also appears to extend to people's behaviour and social observation. In commercial spaces where wood products are used, visitors had more favourable first impressions of staff than in spaces without wood.

One interesting observation is related to housing for the elderly. When wooden materials and wood surfaces began to be used in housing for the elderly, the interaction between them and their interest in the environment increased.

Wood is also anti–bacterial. It has been proved to prevent the growth of dangerous microbes Consequently, wood is used in saunas, washrooms and kitchens.

It has long been known that wood surfaces can affect acoustics and indoor air quality. Traditionally, the acoustic properties of wood have been put to use in instruments, lecture rooms and concert halls.

生理上的数据测量显示，木饰面房间中人的睡眠质量和从紧张环境中的恢复能量都比仿木饰面房间中的要好。

木饰面房间产生的积极生理效应在诸多学校空间的研究数据中也被发现证实。在实木装饰的教室内测试在校学生的脉搏，得到的压力峰值在学生到达教室后不久就消失了。反之，在相应的测试教室空间中，一定的压力情况会持续一整天。在木质饰面教室中学生的压力感、疲劳感和缺乏成就感的体验要比在普通教室中少。

室内使用木材料的这些效应也延伸体现在人们的行为方式和社会调查中。使用木材的商业空间相比非木质饰面的商业空间，来访者对空间中服务人员的第一印象会更好一些。

一个关于老年人住所的有趣的观察发现：当木材和木饰面被用到老年人住所后，老年人之间的交流增多，他们对于周边环境的兴趣也提高了。

木材具有抗菌性。木材已被证实能阻止危险微生物的滋生，从而常使用在桑拿房、卫生间、厨房间中。

很久之前已被证实，木材具有隔音和改善室内空气环境的效应。通常，木材的隔音作用使得木材常被用于乐器房、报告厅和音乐厅（图1-4、图1-5）。

图1-4　Otaniemi chapel, Kaija and Heikki Siren, 1957　奥塔涅米礼拜堂，盖亚，海基西伦，1957

图1-5　Yale Center for British Art, Louis Kahn, 1977　耶鲁英国艺术中心，路易斯·康，1977

Furthermore, wood has the ability to absorb and release moisture, i.e. even out changes in the humidity of indoor air. Steady humidity improves the quality of indoor air and reduces the need for ventilation, which in turn affects the energy–efficiency of the building.

The effects of wood surfaces on the body have been studied in Norway, Austria, Japan and Canada, as well as in Finnish research institutions. The reasons for these positive effects are not yet known, but they emphasize much of the traditional wisdom attached to wood.[2]

From the architectural perspective, however, it is not feasible to build spaces that look too much like saunas, as excessive use of wood may lead to a visual overload.

In the Tampere Hall, we used fibre–cement boards covered with birch veneer. The outcome was acoustically just as successful as it would have been using wooden boards of similar thickness. The environment is a major concern whenever using wood.

Alvar Aalto, who among many other things was a master of wood, recognised the advantages of wood when designing furniture in the 1920s. In many iconic pieces of Modernist furniture, polished steel was favoured (e.g. the Wassily chair by Breuer), whereas Aalto went for bentwood in his designs.

The antibacterial qualities of wood are probably still a fairly unknown territory. However, studies have shown, for example, that the more precious the metal used in door handles, the fewer bacteria they spread.

此外，木材能够吸收空气中的水分，调节室内空气湿度。稳定的室内空气湿度有利于提高室内空气质量，降低通风设备的使用需求，从而提高建筑节能性。

挪威、澳大利亚、日本、加拿大以及芬兰的相关研究机构在建筑的木材表皮方面都展开了研究，产生木材表皮这些正面作用的原因尚不清楚，但研究强调木材使用凝聚着许多前人的智慧（图1-6、图1-7）。[2]

然而，从建筑学的观点看，建造太多看起来像桑拿室的建筑空间并不可行，木材在空间中的过度使用会导致视觉疲劳。

在坦佩雷中心（一个集会议、展览、演出功能于一体的城市公共综合体，位于芬兰坦佩雷市），我们使用水泥纤维板加桦木薄板饰面，这个复合表皮和使用厚度相同的全木板表皮得到了同样成功的声学效果。使用木材料时，环境是主要的考虑要素。

阿尔瓦·阿尔托在20世纪20年代设计家具时，综合了木材的各种优点，相较他所使用的其他各种材料，他是使用木材料的大师。在许多现代家居的典范作品中，光滑的钢材料非常受欢迎（例如布罗伊尔的瓦西里椅），而阿尔托则寻找可弯曲木材来运用在他的设计中（图1-8、图1-9）。

木材的抗菌性可能仍是一个未知领域。然而，研究已表明，例如，门把手使用的金属材料越少，细菌的传播越少。

图1-6　Old Porvoo wooden houses　波尔沃历史小镇的木屋

图1-7　Sea Ranch, Charles Moore etal, 1965　海上牧场 / 摩尔，美国加州查尔斯·摩尔度假别墅（木结构），1965

图1-8　Villa Mairea, Alvar Aalto, 1939　玛利亚别墅，阿尔瓦·阿尔托，1939

Research has been carried out on the role of plants in so-called healing architecture, clearly showing that plants help improve the quality of air both indoors and outdoors.

Wood is a highly versatile building material. Trees felled in the forests are normally cut to different measures and are used as such in building.

Wood can also be made into plywood where layers of thin veneer are glued together, producing boards of varying thickness.

Plywood can be waterproofed, so that it lasts somewhat longer outdoors than ordinary plywood. However, even waterproofed plywood needs to be replaced from time to time.

Wood can also be made into boards by cutting it first into chips and then pressing them into boards with glue. This product is known as chipboard. Wood can also be glued together to make glued laminated timber, or glulam. Glulam can be used even for extremely long spans.

The most recent glulam product to come on the market is CLT panels, which are expected to give wood construction a noticeable boost.

Finland has long traditions in timber construction. Our oldest abodes were built from wood: Lappish huts, pared logs, etc. Wood is a renewable resource, and easy to work.

Japan is another country with a strong tradition of timber-construction. In Japan, the wooden elements of the most significant wooden temples, for example, are renewed every 20 years, so the buildings look as if they are new.

对于植物在所谓"治愈建筑"中作用的研究，清楚地表明其有助于改善室内外的空气质量。

木材是一种用途多样的建筑材料。森林中的树木通常被砍伐切割成建筑使用时需要的模数（尺寸）。

木材可以制作成用一层层薄木板黏合在一起的胶合板，以此生产出各种不同厚度的面板。

防水胶合板在室外的使用寿命长于普通胶合板。然而即便是防水胶合板也需要定期更新替换。

木材也可以通过切碎成小条块用胶水黏合压制成板材，这种板材就是我们熟知的刨花板。木材还可以加工成模板叠合黏压在一起的集成材。集成材可以用在大跨度结构上。

市场上最新出现的集成材产品叫 CLT 板（交错层压木板），人们对于这种板材带来木结构的革命性进步寄予厚望。

芬兰的木结构有久远的传统。芬兰最古老的住宅是木材料建造的：拉普兰小屋（Lappish huts）、原木板小屋（pared logs）等。木材是一种可再生资源，且易于加工。

日本是另外一个传统悠久的木结构建筑国家。例如，在日本，重要的木结构寺庙的木材每 20 年会更新替换一次，因此这些建筑能保持历久弥新的状态。

图 1-9　IIE New York, Alvar Aalto, 1965　纽约国际教育协会主会堂，阿尔瓦·阿尔托，1965

In Southern Europe and around the Mediterranean, forests were all but obliterated at an early stage for housing and ship building. Today, their shorelines are largely treeless. In Southern Europe this has resulted in the disappearance of the timber construction tradition. In China, bamboo structures are becoming a major trend.

The warmth and softness of wood can be sensed with your fingertips. Wood is comfortable to the touch. The surface of wood is porous like human skin. Our sense of touch is in fact essential in how we experience architecture.

Finns love their forest; some prefer spruce and pine, others silver birch. My childhood memories include visions of tall, slender white trunks of young birch tree stands. Climbing up their trunks and the feel of it bending under my weight was a wonderful sensation. The tree giving in gave that lovely sinking feeling in the stomach. I wonder if this is how Alvar Aalto also found his inspiration for the wooden reliefs that he employed in many of the public buildings he designed.

Wood does not withstand natural elements indefinitely. It is sensitive to moisture and will rot. If moisture finds its way indoors, wood will grow mould on its surface. Mould in indoor air is a major health hazard.

Old log buildings were typically covered with board cladding, which could be replaced when it became too dilapidated.

Painting a building with a suitable colour helped it to blend in with its surroundings. History has many examples of wooden buildings which were painted to look just as prestigious as their neighbouring stone buildings. Many designers, however, prefer the natural colour of wood as more genuine. Wood can be treated with colourless impregnation agents which prolong the life–cycle of a wooden structure. This retains the real characteristic of the wood's surface. The beauty of the grain remains visible.

Different varieties of wood are used for different purposes in building; the inherent qualities of a type of wood direct its use.

在欧洲南部和地中海周边地区，早期的森林木材几乎都用于建造住宅和制造船只。今天，这一地区的海岸线几乎成为不毛之地，没有一棵树。这导致现在欧洲南部的木结构建筑几乎消失殆尽。而在中国，木结构曾经是最主要的建筑类型，但现在竹结构建筑正在成为一种趋势。

木材给人指尖的触感是温暖柔和的，摸起来很舒适。木材表面如同人类皮肤一样具有渗透性，事实上我们对建筑的感受很大程度上取决于我们的触感。

芬兰人热爱他们的森林，有些人更喜欢云杉林和松树林，有些人则更喜欢白桦林。我的童年记忆中就有高高细细的小白桦树干的身影仁立着。爬上白桦树把树杆压得向下弯曲，是一种奇妙的感觉。树枝在向下弯曲的过程中让人腹部产生了舒服的下坠感。我一直想知道这是否就是阿尔托所说的木材使人放松，从而在很多的公共建筑设计中使用木材的灵感来源。

木材并不能一直保持自然元素属性。它对于潮湿非常敏感而且有腐烂的可能。如果潮湿找到机会入侵室内，木材表面就会长霉，室内环境中的霉菌是人类健康的主要威胁所在。

老的木建筑通常覆以木板，残破的木板可以替换掉。

用适当颜色装饰建筑有助于建筑与环境的协调。历史上有许多木结构建筑因为适当的漆面而看起来毫不逊色于与其相邻的石材建筑。然而，许多设计师认为原木色更为真实质朴。可以使用防腐清漆（无色浸渍剂）处理木材来延长木结构的使用寿命。这种清漆可以保留木材表面的特征，美丽的木质纹理依旧清晰可见。

不同木材在建筑中有不同的用途：一种木材的内在品质直接决定了这种木材的用途。

In Finland, the most popular wood varieties in building are pine, spruce and birch. Less common species, such a maple, ash, beech, oak and aspen, are also used, particularly for interiors.

Imported varieties, such as teak, ebenholz, jacaranda, mahogany, abachi are used for more special spatial effects or functional quality. Hardwood is durable, while softwood stores heat.

Wood is fairly fragile as a material, but easy to work, even by hand.

With foreign wood even more durable outdoor structures can be achieved. Another point of consideration is the availability of wood varieties, which also dictates which materials are chosen.

The surface of wood can be pared with a broad–axe to achieve a hand–hewn appearance. Log building techniques such as this have long been applied. Timber is most commonly used as boards, with a sawn or planed surface. Sawing leaves a more vibrant surface, although many prefer a smoother finish that prevents splinters in fingers.

Wood burns easily. Many old wood–built towns have burnt to the ground many times over the course of their history. Today, wooden buildings are typically protected against fire with water sprinklers.

Massive glulam structure have a good fire safety, as when they burn they char but do not collapse.

The oldest wooden building in the world still in use is probably the Buddhist temple Horyu–ji, which was completed in 607 AD. The central column of the temple, made from cypress, was felled in 594. It is part of Japanese wood construction culture that wooden parts are replaced with new ones from time to time, when necessary.

在芬兰，建筑中最常用的木材是松木、云杉木和桦木。用得较少的品种如枫木、橡木、白蜡木、山毛榉木、橡木和山杨木，主要用于室内。

进口木材品种，如柚木、乌木、檀木、桃花心木、阿巴奇木主要用在一些特殊空间或特殊功能空间。硬木耐久而软木保温性能好。

木材是一种相当脆弱的材料，但易于加工制作，甚至可以徒手制作。

进口木材可以制作耐久的室外结构。另一个考虑要点是木材品种的可用性（原材料树木的数量），这也决定了材料的选择。

木材表面可以用斧子削出手工雕琢的外观，木屋建造一直以来都运用这项技术。人们常常通过对木材表面的锯平或刨平加工制成板材来使用木材。锯出来的表面更有质感，但是许多人更喜欢不扎手的光滑表面。

木材易燃，许多木材建造的小镇在历史发展中不止一次被烧成平地。如今，木建筑通常都装有消防喷头来预防火灾。

使用集成材的建筑都有很好的防火安全性，一旦着火，集成材会碳化，但不会整体倒塌。

世界上最古老的仍在使用中的木结构建筑可能是建成于公元607年的佛教寺庙法隆寺。建造寺庙的中心柏木柱砍伐于公元594年。在必要时不断替换更新建筑木组件是日本木建筑文化的重要组成部分。

Europe's oldest wooden building is the Greensted Church in the UK. Its vertical oak trunks were felled in the late 11th century AD.

The oldest wooden residential house is the Bethlehem House in Switzerland, built in 1287.

Matter retains its own logic. Organic materials are rare homogeneous; rates of expansion and contraction differ depending on orientation in all directions, and stress and tension is tolerated in varying fashions. The way a material is shaped also carries elements of tradition, defined by the available processing methods.

Wood has excellent material properties; it is exceedingly tactile; and both versatile and simple to work. Its load–bearing and insulating properties mean that, in a northern climate, it is possible to leave load–bearing structures uncovered.

In fact, this makes it the only appropriate choice of material in Finland when pursuing a Tectonic–Constructivist approach to architectural design.[3]

Wood is an organic material. When used sustainably, it is a renewable natural resource. Wood is said to be environmentally friendly, but in scientific terms this can only be verified based on a value analysis applying the input–output model on different materials in otherwise comparable situations. However, it is known that trees do absorb carbon and when used as products wood serves as carbon storage.

Wood is soft, familiar and, when correctly used, safe. Wood is durable but has a high charring rate.

Wood teaches us about nature. It carries with it the dizzying memory of freely blowing winds. Wood also teaches us about modesty. Wood is the material of poor and rural people.

Wood is the material of northern people. The rich city folk build from stone. As

欧洲最古老的木建筑是英国的格林斯特德教堂。教堂里的橡木柱在公元11世纪末倒塌。

最古老的木结构居住建筑是1287年建造的瑞士伯利恒之家。

任何事物都有自身的逻辑。有机材料很少是匀质的；收缩膨胀率会随着材料构成物质方向的不同而不同，而抗压和拉升的弹性取决于不同的形态。材料的成型方式带有传统元素的影响，并取决于可用的加工制作方法。

木材具有良好的材料特性、触感好、功能多样、易于加工处理。木材良好的承重和保温的双重特性非常重要，北方的气候使得将承重木结构裸露在外成为可能。

事实上，这些特性使得在芬兰木材成为寻求成功的构成主义建筑设计的唯一的合适材料。[3]

木材是一种有机材料。可持续地利用木材，它能成为可再生的自然资源。木材是环保材料，但是从科学层面来讲，这只是建立在不同材料的投入产出模型价值分析或类似的情况下才成立。然而，树木吸收碳是确凿无疑的，木材作为产品可以用于碳储存。

木材是一种柔软亲和且在正确使用的前提下相当安全的材料。木材耐久性好但碳化率高。

木材教我们认识自然。木材唤起我们自由呼吸和清风拂面的美妙记忆。木材教会我们谦逊。

木材是北方人民的建筑材料。富裕城市的人们用石材建造。如地中海地区的人们，因为他们的森林在很久以前就被砍伐殆尽了。短暂性、

do the folks in the Mediterranean, whose forests have been cut down long ago. Temporality, transience and rebirth are all part of the story of wood.

Wood is vernacular, robust, simple and warm. Wood sourced from a tree that was allowed to grow slowly, cut to appropriate measurement and tarred, could last as shingle roofing for centuries.[4]

In the processing industry, pulp is produced from wood to be used especially in manufacturing paper, paperboard and cardboard.

Paper has been used in construction in many ways. The partitions in Japanese houses have traditionally been made of paper (rice paper). In Finland, the interior walls of wood–built houses were typically decorated with wallpaper and the ceilings fitted with board and painted. Board is also used in buildings as a windscreen.

Kobe earthquake houses were built from round cardboard tubes in 1995 (architect Shigeru Ban).

Materials and surfaces surely have a language of their own. Stone speaks of its distant geological origins, its durability and inherent permanence. Brick makes one think of earth and fire, gravity and the ageless traditions of construction. Bronze evokes the extreme heat of its manufacture, the ancient processes of casting, and the passage of time as measured by its patina. Wood speaks its two existences and timescales; its first life as a growing tree and second as a human artefact made by the caring hand of a carpenter or cabinetmaker. These are all materials and surfaces that speak pleasurable of material metamorphoses and layered time.

Juhani Pallasmaa: *The Enbodied Image*

时效性和可再生性都是木材广为流传的特性。

木材具有地域性、简洁粗放和保温的特性。作为木材来源的树木可以慢慢生长，人们从树上裁剪下合适尺寸的木材，涂上桐油，可以使之成为使用达数世纪之久的木屋顶。

在加工产业中，用木材生产的纸浆被用于生产纸、纸板和卡纸。[4]

纸在建筑上有多种用途。日本民居建筑中的隔墙历来是用纸（宣纸）做的。在芬兰，民用木建筑内墙通常用墙纸做饰面，天花板则用木板刷漆。这种木板也可用作室内的屏风。

1995 年神户的抗震民居建筑是用纸管建成的（建筑师坂茂）。

建筑材料与表皮有自己的形式语言。石头讲述他们久远的地质学来源，石头天生具有恒久性。砖块让人想到土地与火，万有引力和建筑结构一直以来的传统。青铜让人想起古代铸造产业中铸铜的极度高温，其上的铜绿是时间旅程的度量衡。木材讲述着两种存在形式和不同的时间维度；它的第一种存在形式是生长的树木，第二种是由木匠或木加工者灵巧双手人工制造的木产品。这些木加工材料和木家具都讲述着令人愉快的材料变化和时间层积。

尤哈尼·帕拉斯玛《形象的表现》

CLT (Cross Laminated Timber)

The most recent arrival to the wood materials market is cross–laminated timber, or CLT. It is promoted as the new concrete. Wood constructors emphasise the importance of calculating a buildings carbon footprint when granting building permits.

CLT is a material in which wood boards are glued across each other in layers. Thanks to the cross–laminated structure, the material does not shrink and swell like wooden structures typically do. Thanks to its stiffening properties, CLT is particularly practical in multi–storey buildings, but is equally suitable for low–rise houses and, for example, bridges.

CLT is a renewable, environmentally sustainable material that guarantees a good indoor air quality, is fire–resistant, has good insulation properties and acoustics. It is almost like a modern equivalent of the log.

According to research, the carbon footprint of CLT is larger than that of steel. The carbon footprint of the log is one third of that of CLT. There is currently insufficient research data on the risk of mold caused by the glue in CLT.

There is currently a factory under construction in Finland, which can substantially improve the potential of wooden multi–story buildings.

When the modules are manufactured to finish at the factory, they are at no stage exposed to moisture. CLT buildings do not require a plastic membrane as a vapour barrier, as the panel has the capacity to level out moisture, like a log wall.

CLT board is versatile, making timber a fascinating construction material. It offers opportunities that concrete, logs or substructures cannot match. Its basic properties include a natural moisture barrier, a rigid structure and dimensional accuracy. The joints are tight, and the boards are quick and easy to fix.

CLT（交错层压木板）

木材料市场最新的产品是交错层压木板，即 CLT。它正在成为一种新型的"混凝土"材料——坚固、抗震、环保。木结构建筑在申请建筑许可时会强调建筑的碳足迹（环保性）。

CLT 是一种用实木板层层方正交叠胶合而成的板材。这种交叠黏合的结构使得 CLT 不会像普通木材结构那样收缩膨胀。而由于 CLT 的硬度特性，这种材料可完全胜任多层高层建筑结构，但同时也适用于低层的房屋或者其他如桥梁这一类的构筑物。

CLT 是可再生、环保可持续的建筑材料，是室内良好空气质量的保障，它具有防火、保温、吸音等各种良好特性。CLT 几乎就是现代版的原木。

研究表明，CLT 的碳排放量大于钢材。而原木的碳排量仅为 CLT 的三分之一。但目前并没有充足的研究数据表明是 CLT 加工过程中的黏合胶是否会导致某些风险。

在芬兰有一个在建的工厂，它可以表明 CLT 在多层建筑上的发展潜力。

当 CLT 模块在工厂组装完成后，它们无论何时都不会受到潮湿的威胁，CLT 不需要塑料膜作为防水外膜，因为 CLT 的板面能平衡本身的湿度，如同原木墙一样。

CLT 用途多样，它使得木材成为非常棒的建筑材料。它有混凝土结构、原木结构或者基础结构无法比拟的各种优势。CLT 的基本特性包括天然防潮，刚性结构和尺寸精确，模块间结合紧密，面层能迅速便捷地安装等。

Horizontal structures can include projections, canopies and stair solutions.

Indoors, CLT boards can be left visible, and they can be used in all structures above the foundations. Exteriors can be clad with timber, plaster insulation or plaster panels. CLT can be combined with steel and glass because of its dimensional accuracy.

For homeowners, the material offers energy efficiency, safety and comfort. For example VTT Technical Research Centre of Finland and Holzforschung Austria have both found CLT's fire resistance qualities to be outstanding. In order for wood to catch fire, the water it contains must first evaporate. The surface of the panel slowly chars, and the charred layer protects the inner layers, preventing the massive wood structure from collapsing.

Stora Enso is the largest manufacturer of CLT boards in the world. Over the years, experience has been accumulated on small houses as well as large, architecturally demanding private and public buildings.[5]

Thermowood

Thermowood is highly suitable as an exterior material. During the thermal treatment of the wood, its equilibrium moisture content decreases, so thermowood does not react to variation in humidity as much as non−treated wood does. Therefore the dimensions and shape of thermowood are significantly more stable than those of traditionally manufactured wood products.

Untreated wooden surfaces age beautifully acquiring a subtle grey sheen.

CLT 板能提供水平结构，包括出挑结构、檐篷（雨篷）和楼梯等的解决方案。

在室内，CLT 板可以裸露在外，而且可以用于基础结构之上的所有结构。饰面层一般可以用木材、保温石膏或者石膏板。因为 CLT 可以精确地调整尺寸，所以 CLT 可以和钢材、玻璃结合使用。

对屋主而言，建筑材料要节能、安全而且舒适。例如，芬兰超薄技术研究中心（VTT）和澳大利亚木材研究中心都发现 CLT 具有非常卓越的防火性。木材着火，必须先蒸发掉木材中的水分。CLT 的面层着火后碳化非常缓慢，碳化的表面可保护内部木板层，从而阻止木结构建筑的整体性倒塌。

斯道拉恩索是世界上最大的 CLT 板加工厂。经过多年实践，积累了小型住宅、大型功能性建筑和公共建筑的建造经验。[5]

热处理木材（炭化木）

炭化木非常适合用作建筑的表皮材料。木材在高温热处理过程中，炭化木的含水平衡率降低，因此炭化木性质稳定，不会像普通木材那样因空气湿度变化而产生变化。因此炭化木的尺寸和形态较其他传统木加工产品要更稳定。

不需要做其他处理的炭化木表面年轮有一种微妙的灰色光泽。

Chapter 2 Bendable as bamboo

Bamboo is the fastest growing tree variety. It is also said to be an environmentally friendly building material. It has been particularly popular in Asia since time immemorial.

The Italian architect Paolo Soleri was one of the first to draw architects' interest to bamboo construction.

Bamboo is an extremely durable, yet flexible material. It is also widely used in scaffolding. Only now that techniques for cross–laminating and pressing bamboo have been developed, its Renaissance is beginning to look likely. It can be used for building strong structures similar to glulam, in pillars and beams. Of late, it has also gained foothold as an interior material; examples include the Barajas airport terminal in Madrid (Richard Rogers) or Wuxi Grand Theatre in China (Pekka Salminen).

The word "bamboo" was introduced by Carl von Linné in 1753. Bamboo is a grass plant like rice, corn and sugar cane. Different to these, the lignin tissues becomes after some years a structure as hard as wood, but more flexible and light. Bamboos, in their wild form, grow on all of the continents except Europe. (see map). There are tropical and subtropical bamboos that thrive in different ecological niches. The majority of species are found in warm zones with humidity levels of over 80%, in tropical cloud

第二章　柔韧之竹

竹子是一种生长迅速的植物，它也被认为是一种环保材料。在亚洲，竹子历来是一种特别流行的建筑材料（图 2–1）。

意大利建筑师保罗·索勒是最早对竹结构建筑产生研究兴趣的建筑师之一。

竹子是一种非常耐用而柔韧的材料，也常用于脚手架。

现在交叉压层和竹板压制技术才刚刚起步，竹结构的复兴即将兴起。竹子可以用作柱子和梁，建造与胶合板相媲美的坚固的建筑结构。近来，竹材在室内材料中开始有立足之地，如西班牙马德里巴拉哈斯国际机场航站楼（理查德·罗杰斯）（图 2–2），中国无锡大剧院（佩卡·萨米宁）。

竹子这个词是 1753 年由卡尔·林奈引进欧洲的。竹子，如同稻子、玉米、甘蔗一样是一种禾本科的植物。与其他几种禾本科植物的不同之处是，经过数年生长后，竹子的木质部会变得和树木一样坚硬，但是又更具柔韧性、更轻。野生的竹子生长在除欧洲之外的世界各洲。热带竹、亚热带竹在各自适合的不同生态位茁壮生长着。人们发现大部分竹子品种都生活在湿度高于 80% 的温带地区，一般是日照少的热

forests, and in clayey and humid soils. It is estimated that 37 million hectares are covered with bamboo forests. [6]

Since antiquity, bamboo has been a construction material used to build basic habitats to complex structure; it has formed part of a set of elements that were an essential part of cultural development in Asia and America.

Bamboo is a rapid growing natural resource that can produce much more dry biomass per hectare per year than eucalyptus. The production of bamboo biomass depends on many factors and therefore varies significantly.

Plants that assimilate CO_2 for photosynthesis, storing it in their biomass, make an important contribution to the global climate. Because of its rapid growth, bamboo can take in more CO_2 than a tree.

Due to its favourable mechanical characteristics, great flexibility, rapid growth, low weight and low cost, bamboo is a construction material with many applications. It is estimated that one billion people live in houses constructed from bamboo. In seismic zones bamboo construction is preferred due to its lightness and flexibility.

The majority of traditional houses in the rural zones of warm humid climates where bamboo grows, are constructed of raw bamboo. A typical use of bamboo canes is in the construction of scaffolding. In Asia these are found with heights of more than 40 storeys.

Because bamboo members are hollow, they represent a high fire risk. Nevertheless, the external layer of the bamboo canes contain a high concentration of silicates and are therefore not highly flammable.

Asia has pioneered the industrial development of the use bamboo in laminates.

带林区，潮湿的黏土状土壤环境下。现已证实有
3700 万公顷的土地被竹林覆盖（图 2-3）。[6]

自古以来，从基本的栖息构筑物到复杂结构的建
筑，竹子是一种常用的建筑材料；在亚洲和美洲
的文化发展中，竹子逐步成为一套完整建筑元素
体系中的重要组成部分。

竹子是一种生长迅速的自然资源，每公顷每年能
生产的干生物量（干重[1]）比桉树更多。竹子每年的
单位面积生物量的产出因各种决定因素而有显著
变化。

植物吸收二氧化碳进行光合作用，将其存储在生
物量中，为全球气候做出重要贡献。由于生长迅
速，竹子能比树木吸收更多的二氧化碳。

竹子良好的力学特性，极强的柔韧性，生长迅速，
质量轻和低成本，使它成为一种用途广泛的基建
材料。竹竿一个典型的用途是搭脚手架。在亚洲，
竹竿可以搭出 40 层以上高度的脚手架。

由于竹竿是中空的，这意味着火灾风险高。然而，
竹竿表层较高的硅酸盐含量又降低了其可燃性。

亚洲率先在层压板材行业中使用竹子作为原料。

图 2-1　Bamboo world　世界产竹
区地图

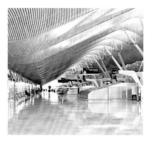

图 2-2　Barajas Terminal, Madrid,
Richard Rogers，2004　西班牙马
德里巴拉哈斯国际机场航站楼，理查
德·罗杰斯，2004

图 2-3　Bamboo forest　竹海

1　干重：在生物学上指细胞除去自由水后的重量干重。干重是
　　指把自由水去掉，剩下的是脂肪、蛋白质、维生素、无机盐
　　等营养物质。（译者注）

Bamboo laminates can be used as an alternative construction material. For laminated elements such as planks, boards, columns, beams and panels, laths joined with glue or wooden are used. To produce laminates, the interior and exterior of the strip must be cut , from which pieces with rectangular sections are obtained and joined.

Panels of laminated bamboo were developed in China 1982. The strips are glued with phenol formaldehyde or urea formaldehyde orthogonally in an alternating manner. The pieces can also be united with the help of heat and high pressure. Building with bamboo laminates. We need more knowledge and construction details in joints in the future. Also the gluing needs to be examined more carefully, because of the possible mold.

In interiors and furnishings, bamboo has been rediscovered and regained popularity in Asia. It is a growing trend to build more with bamboo laminates today.

竹材层压板可用作替代性的基建材料。层压材料元素如厚板、薄板、长柱、短柱、板条等用胶水黏合。这些黏合的坯条必须内外切平，从而得到或组合形成矩形截面的层压材料。

1982 年中国发展出了竹材层压板。竹篾用苯酚甲醛或者尿素甲醛作为黏合剂，采用正交方式进行黏合。这些竹篾黏合条通过高温高压（热压）加工形成竹材层压板。

未来在建筑上使用竹材层压板还需要了解更多的相关知识和研究接缝施工详细的工艺。当然胶合剂也需要进行进一步的检测，因为它会对竹材层压板的性能产生影响。

在室内设计和家具设计业，竹材被重新认识发现并再一次在亚洲流行起来。如今，用竹材层压板建造已成为一种流行趋势（图 2-4）。

图2-4 Tampere Hall, Sakari Aartelo and Esa Piironen, 1990 坦佩雷音乐厅萨卡利·阿尔泰洛、埃萨·皮罗宁, 1990

Chapter 3   Hard as stone

Material has the effect of making
unity. All art forms are based on
matter; they have to confront
materiality.

Alvar Aalto

Stone is a material that contains minerals and is classified based on their mineral content.

It is divided into igneous, sedimentary and metamorphic rocks, depending on the origin. Igneous rocks are formed by solidified magma, while sedimentary rocks are formed by the deposition and subsequent cementation of older rocks and metamorphic rocks are a result of igneous and sedimentary rocks being subjected to very high heat and pressure deep inside the earth's crust.

Stone is an inorganic natural material. Objects made from stone can be preserved in the soil for thousands of years without degrading. Stone is the symbol of sustainable building.

Rock caves have provided shelter for people and animals since time immemorial. Caves bring various images into minds. Stone is also one of the original natural materials, with a long history as a building material. Perhaps the most impressive historical monuments are the Egyptian pyramids, which are still standing today.

第三章　坚硬之石

材料具有完整性效应。所有的艺术形式都以物质为基础，都必须面对物质材料。

——阿尔瓦·阿尔托

石材是一种含有矿物成分的建筑材料，常以矿物成分来进行石材的分类。

根据来源分类，石材一般分为火成岩、沉积岩和变质岩。火成岩由岩浆凝固形成，沉积岩是由原来的石头层层沉积、石化积淀而成，而变质岩则是火成岩和沉积岩在地壳深处受高温高压作用后形成的。

石头是天然无机材料。石头制品可以在土壤中保存数千年而不变质。石材是可持续建筑的象征。

远古时代，石洞就已经给人类和动物提供庇护场所。人们对于洞穴有各种各样的想象。长期以来，石头作为建筑材料历史久远，它是最早的天然材料之一。可能令人印象最为深刻的是如今依然伫立着的埃及金字塔（图 3-1）。

图 3-1　Great pyramid Giza, Hemiunu 2560 BCE　吉萨大金字塔（胡夫金字塔），赫米翁努 公元前 2560

图 3-2　Hatsepsut Temple, Senenmut 1500 BCE　哈特谢普苏特神庙，森内马特 公元前 1500

图 3-3　Luxor　卢克索神庙

It is still not known quite how these magnificent constructs were built. Whatever the method, it was extremely labour intensive. The pyramids were built as burial places for the Pharaohs, and they are shrouded in legends and beliefs that have only gradually come to light in the modern age.

Stone was used for erecting fortifications and fortresses against enemies. Wealthy religious communities built places of worship from stone.

The Ancient Greeks created exquisite architectural feats from stone. The ruins on the Acropolis in Athens are an example of this, known throughout the world. Greek stone architecture is divided into three epochs, Dorian, Ionian and Corinthian, manifested in the different designs of columns.

The Romans developed stone construction a step further from the column and beam constructions and introduced arches. The Romans also went on to create the rudiments of concrete construction alongside stone; they invented composite structures that were reinforced with plaster.

Stone is a highly usable material in construction, as there is plenty of it in the soil and it is durable. In some areas, stone quarries have already been depleted, but new quarries can be opened elsewhere, and stone is shipped around the world today. For example, Finland exports large quantities of raw stone to China and Italy. The product development of stone has progressed significantly with improved industrial methods. Earlier, stone was quarried mainly through very simple means.

With the development of explosives, blasting stone from the bedrock became easier. Technical tools have replaced manual stone cutting.

The hardness or softness of a stone strongly determines its application in building. In Southern Europe, softer stones, such as marble and sandstone, are the main types of stones used.

我们至今仍不知道这些宏伟的建筑是如何建成的。无论采用何种方法，都需要极为密集、高强度的劳动。金字塔是法老们的墓穴，一直笼罩在各种传说和信仰中，直到现代才逐渐被人们了解。

石材被用来建造防御外敌的工事和堡垒。富有的宗教团体会用石材建造膜拜广场（图3-2～图3-4）。

古希腊人创造了精湛的石头建造技艺。世界闻名的希腊雅典卫城遗迹就是一个范例（图3-5）。希腊石建筑根据柱式不同分为三个时期：多立克、爱奥尼和科林斯。

从梁和柱式结构到拱门结构，罗马人进一步发展了石结构。与此同时，罗马人创造了混凝土施工的雏形，他们发明了用灰泥加固的复合结构。

由于资源充足而且耐久，石头是建筑中非常有用的材料。有些地方的采石场已经枯竭，但又会有新的采石场在新的地方出现。现今，石头已运到世界各个地方，如芬兰出口大量的生石料到中国和意大利。随着工业方法的改进，石材产品的开发力度很大（图3-6、图3-7）。早期的石材开采方法非常简单。

随着炸药的发展，从基岩上爆破石头变得简单很多。机械工具也取代了手工切割石头。

石头的软硬度是决定这一石头材料在建筑中所起作用的主要因素（图3-8）。在南欧，软性的石头，如大理石和砂岩是当地所使用石材的主要类型。

图 3-4　Machu Picchu　马丘比丘，秘鲁

图 3-5　Parthenon, Athens, Iktinos and Kallikrates, 438 BCE　帕特农神庙/万神庙，雅典，伊克提诺斯和卡拉克拉特，公元前438

图 3-6　Yale Rare Book Library, Gordon Bunshaft, 1963　耶鲁大学图书馆，戈登·邦沙夫特，1963

图 3-7　Whitney Museum, NYC, Marcel Breuer, 1963　惠特尼博物馆，纽约，马塞尔·布劳耶，1963

图 3-8　Temppeliiaukio Church, Helsinki Timo and Tuomo Suomalainen, 1969　岩石教堂（坦佩利奥基奥教堂），添姆和杜姆苏马连宁兄弟，芬兰赫尔辛基，坦佩雷，1969

Carrara marble is a highly regarded stone variety, used in statues and sculpture and high prestige buildings.

Finlandia Hall in Helsinki, designed by Alvar Aalto, is one of the greatest examples of Carrara marble clad buildings of our time in the world.

The Finnish climate has also revealed the downside of building from this particular material. The concert hall is clad with slabs that are too large and too thin. In addition, the design left an insufficient ventilation gap behind the slabs, with the result that the stones began to bend after 30 years.

All the slabs were replaced with the same, famous Carrara marble in the early 2000s. It is already evident that the slabs need to be replaced again in the future. This cycle of replacement is similar to that in Japanese temples, where wooden elements are replaced every 20 years.

Despite the shortcomings in the design of the Finlandia Hall (its acoustics were also unsuccessful), Alvar Aalto's architecture is highly reputed internationally.

Stone varieties of Finnish origin are harder and more durable than marble. Grey granite will last for a hundred years in the Finnish climate without signs of deterioration. A major threat to stone materials comes in the form of pollution. Finland is also the origin for rare soapstone, which, thanks to its porosity, is ideal for heat–storing fireplaces and stoves.

Stone was earlier used in loadbearing structures, but today, it is only used for external cladding. Stone slabs are thin and typically secured onto the frame with metal attachments. Stone slab seams can be sealed with different types of plaster or weatherproof putty and fillers or they can be left open.

白色大理石是一种美誉度极高的石材类型，主要用于雕塑、雕刻和高贵的建筑。

阿尔瓦·阿尔托设计的赫尔辛基芬兰宫，是我们这一时代白色大理石饰面最伟大的建筑之一（图3-9）。

图3-9　Finlandia Hall, Helsinki, Alvar Aalto, 1974　芬兰宫 阿尔瓦·阿尔托，1974

然而，芬兰的气候揭示了使用这一材料建造建筑的负面效果：音乐厅所使用的大理石面板过大过薄，加上设计中保留在面板后面局促的通风间隙，导致在30年后这些大理石开始弯曲。

图3-10　Essen Opera House, Alvar Aalto, 1985　埃森歌剧院，阿尔瓦·阿尔托，德国北莱茵威斯特法伦州

进入21世纪，所有的大理石面板都换成了白色大理石，显然，未来还会再被替换。这种更新周期与日本寺庙相似，日本寺庙的木材每20年更换一次。

图3-11　American Center, Paris, Frank Gehry, 1994　美国中心，巴黎，弗兰克盖里，1994

尽管芬兰宫（一个音乐厅）有声效设计方面的缺欠，但阿尔瓦·阿尔托的建筑（图3-10）仍在世界上享有盛誉。

芬兰原产的石材要比大理石坚固且耐久。在芬兰的气候环境下，灰色花岗岩能保持百年以上没有侵蚀的迹象。石材的主要威胁来自污染。芬兰也出产稀有的鸡血石，它的多孔性，使它成为一种储热壁炉和火炉的最佳建造材料。

图3-12　AT&T NYC, Philip Johnson, 1984　纽约（AT&T）大厦，飞利浦·约翰逊，1984

石材最早用作承重结构，如今多用作建筑表皮材料（图3-11、图3-12）。薄薄的石材板通常贴附在金属结构的框架外面。石材间的缝隙可以用水泥或防水腻子填补或者直接保留缝隙。

In the northern climate, it is necessary to leave a gap between the stone cladding and the building frame to allow the weatherproof shell of the building to breathe.

When correctly built, stone structures are a durable choice. Stone is also a fireproof material, and therefore safe. The surface of stone can be treated in many different ways. Blocks of stone quarried from rock were traditionally used as they came either in building foundations or the façade. As kerbstones, the surface of stone blocks is usually chiselled to a rough finish. For a more even finish, the surface can be bush hammered, a process that was traditionally also carried out with chisels. This type of stone can be used for floors and façades. The surface is only slightly coarse. If an even smoother finish is required, the surface can be honed or even polished. This of course means more work, and hence a higher price tag. Sometimes stone floors in public spaces are worked to a finish that is too smooth, making the floor dangerously slippery even with a small amount of water or dirt.

Stone is an eternal natural material. When you pick up a stone on a beach, you are holding billions of years of history in your hand.

The pleasant memories of feeling is the surface of a rock, warm in the sun, after a swim, a memory that many of us hold dear.

"Natural materials–stone, brick and wood–allow the gaze to penetrate their surfaces and they enable us to become convinced of the veracity of matter. Natural material expresses its age and history as well as the tale of its birth and human use. The patina of wear adds the enriching experience of time; matter exists in the continuum of time. But the materials of today–sheets of glass, enamel metal and synthetic materials–present their unyielding surfaces to the eye without conveying anything of their material essence or age."

Juhani Pallasmaa: An Architecture of the Seven Senses

在北方的气候环境下，石材幕墙和建筑框架之间必须适当留缝以保证建筑防水层的呼吸（膨胀与收缩）。

当采用正确的方法去建造时，石材结构是一种经久耐用的选择。石材防火，因而使建筑具有安全性。石材表面可以用各种方式去处理。从岩石场开采来的石材通常要么用于建筑基础要么用于建筑表面。作为收边石，石材的表面一般被雕凿成毛面。略为平坦的毛面一般是荔枝面，这一过程一般是用凿子凿出来的。这类石材可以用作建筑地面或者墙面。荔枝面类的石材面层稍有些粗糙。如果是需要更加光滑一点的面层，则可以进行面层的磨光甚至抛光。这意味着更大的工作量，因此价格更高。有时候公共空间的石材地面由于太过光滑，稍有水或灰尘，人们就有滑倒的危险。

石材是一种永恒的自然材料，当你在沙滩上捡起一块石头，上亿年的历史就握在你手中。

我们中的许多人都拥有这样的美妙回忆：在游泳之后，我们可感受到在阳光下触摸岩石时的温暖。

石头、砖、木材，我们的目光穿透这些自然材料的表面，它们使我们确信物质的真实性。自然材料讲述着它们的年代和历史，以及它们的诞生和人类运用它们的故事。材料的包浆强烈喻示着它们在岁月流逝中的多彩经历；物质存在于不断向前的岁月中。但是今天的材料——玻璃板、搪瓷金属和合成材料——它们所呈现的僵硬表面却不再传达任何关于它们本质与年代的信息。

尤哈尼·帕拉斯玛《建筑七感》

Chapter 4 Solid as brick

You should understand the structure of wood, the
weight and hardness of stone, the character of
glass; you should become one with your materials
and shape and use them in accordance with their
nature. If you understand the nature of a material,
you become intimate with its potentials in a far
more tangible way than through mathematical
formulas and constants.

Jorn Utzon

Brick is a ceramic building material made from clay and sand by curing or firing
to a regularly shaped and sized unit. The typical profile is a rectangular block.
Holes can be added to bricks before firing, which improves their compressive
strength. To the clay mass can also be added sawdust, which increases the
porosity created during firing and thereby the brick's frost resistance.

Sand–lime bricks are not fired, and instead are made of slaked lime and quartz
sand, pressed into shape. The bricks are not fired, as they are steam cured. The
more correct term would, in fact, be sand–lime stone.

The first brick buildings were built in Uri, Mesopotamia, c. 4000 BC. Bricks
were also manufactured in Egypt as early as in 3000 BC. These bricks were made
from clay, but they were not fired, only cured (clay brick). Mesopotamians had
learnt the art of firing bricks by 2200 BC.

Among the most notable early buildings made from brick are the Great Wall of

第四章　方正之砖

你应理解木材的构造、石材的坚硬与重量、玻璃的特性；你需成为根据材料天性来塑造和运用它们的人。若你理解了材料的天性，你就能用更为具体的方法而不只是用数学公式计算出来的方法来挖掘材料的潜力。

——约翰·伍重

砖是一种由黏土和沙子制成的陶瓷建筑材料，通过固化或烧制形成规则形状和一定大小的单体。最典型的形状是矩形块。烧制前可以在砖上增加孔洞，以此来提高砖块的抗压强度。也可以在黏土原料中增加木屑，可以在烧制过程中增加孔隙从而提高砖块的抗冻性。

灰砂砖不经过烧制工序，而是使用熟石灰和石英砂压制成形，这些砖坯不经烧制，而是经蒸汽固化。事实上其更为确切的名称是灰砂石。

最早的砖结构建筑建于公元前 4000 年的美索不达米亚平原乌尔地区。早在公元前 3000 年，埃及也开始制造砖。这些砖由黏土制成，这些砖也不经烧制，而仅是固化的黏土砖。美索不达米亚人到公元前 2200 年左右学会了烧砖技艺。

世界上早期著名的用砖建造的建筑有中国的长城（公元前 210 年）和

图4-1　Rautavaara Church, Sakari Aartelo and Esa Piironen, 1982　劳塔瓦拉教堂，萨卡利·阿尔泰洛，埃萨·皮罗宁，1982

China (c. 210 BC) and the Pantheon in Rome (c. 128 AD). Brick architecture spread to northern Europe with Hanseatic trading and the Teutonic Order in the 12th century.

From the Middle Ages up until the early 1900s, brick structures were solid structures, with the thickness of the walls at least that of the brick. Bonds were created laying headers and stretchers to create stability and stop the wall from falling apart in the middle. One of the oldest bonds was the monk bond, where two stretchers are followed by a header, repeated on the same layer.

Today, brickwork is laid using various running bonds. The most common one is the stretcher bond, the most common variations being the standard or raking stretcher bond, with each successive course staggered by a half and 1/3 of a stretcher, respectively. The most common brick size used in Finland is $257 \times 123 \times 57$.

The first bricks were made from compacted soil. A mixture of clay and straw was shaped into bricks in moulds and dried.

The skill of firing originated in pottery. Clay and bitumen were used in bricklaying.

Builders started to add decorations to brick surfaces as early as 3,000 years ago through glazing.

罗马的万神庙（公元 128 年）。12 世纪，由于汉萨贸易同盟[1]和条顿骑士团[2]，砖结构建筑逐步流传到了北欧（图 4-1～图 4-4）。

从中世纪到 20 世纪初，砖结构一直是实心结构，墙体厚度即砖的厚度之和。人们创造了砖块的顺丁砌法，保证了墙面的稳固，从而防止墙从中间倒塌。最古老的砌法叫梅花丁砌法，即两顺一丁的砌法在同一层上不断重复。

现在，砖块砌筑有各种不同的顺砖压缝砌法。最常见的一种是砖的长边平行墙面的顺砌方法。一般标准的压缝变化操作是错缝 1/2 或者 1/3 的砖长依次重复排列顺砌。芬兰常用的砖的尺寸是257mm×123mm×57mm。

最早的砖是由土压制而成的，由泥土和稻草混合放在模具中压制风干而成。

烧砖技术源于烧陶，黏土和沥青用于砌砖。

早在约 3000 年前，建造者就使用上釉的方式来装饰美化砖块表面。

图 4-2 Robie House, Frank Lloyd Wright, 1910 罗宾之家，弗兰克·劳埃德·赖特，1910

图 4-3 MIT Dormitory, Alvar Aalto, 1948 麻省理工学生宿舍，阿尔瓦·阿尔托，1948

图 4-4 MIT Dormitory, Alvar Aalto, 1948 麻省理工学生宿舍，阿尔瓦·阿尔托，1948

图 4-5 MIT Dormitory, Alvar Aalto, 1948 麻省理工学生宿舍，阿尔瓦·阿尔托，1948

---

1 汉萨贸易同盟（Hanseatic trading）：德意志北部城市之间形成的商业、政治联盟。13 世纪逐渐形成，14 世纪达到兴盛15 世纪转衰，最终解体。（译者注）
2 条顿骑士团（Teutonic Order）：12 世纪末建立的天主教骑士的组织，从抵御异教入侵发展为进行向东欧的军事入侵和移民活动，逐步壮大。（译者注）

Bricks can be used for loadbearing structures and as decorative material enhancing the architecture.

The Romans developed the technique of building arches to new levels. With the Romans, the skills of brick–making and building arches spread to the outermost reaches of Europe. Brick remained the key building material of large churches for a long time. The current size of the brick, which comfortably fits into the hand, was developed in the 1100s.

Around 1140s, the earlier Romanesque rounded arches gave way to the pointed Gothic arch, marking the height of brick construction technique.

Most commonly, bricks were used as decorations in natural stone structures. The decorative patterns were created with shaped bricks and varied bonding.

Brick gradually replaced natural stone as the favoured material in building frames in the 16th and 17th centuries.

To begin with, bricks were made on–site, until brick factories were established from 18th century onwards.

Machine–made bricks were invented in the 19th century, which started a whole new boom in brick architecture.

砖块可以用于建筑承重结构，也可用作装饰材料来提升建筑形式美。

图4-6 IIT Chapel, Mies Van der Rohe, 1951 伊利诺伊理工学院克朗小教堂，密斯·凡·德罗，1951

罗马人将建筑拱券技术发展到了新的水平。罗马人把制砖技术和建筑拱券技术传播到了欧洲的每个角落。砖作为大型教堂建筑的主要建造材料持续了相当长的一段时间。现在所用的砖的尺寸非常符合人手尺度，是12世纪时发展起来的。

图4-7 Säynätsalo Town Hall, Alvar Aalto, 1952 珊纳特赛罗市政厅，阿尔瓦·阿尔托，1952

大约在12世纪40年代，尖尖的哥特式拱门取代了早期的罗马圆拱门，砖结构技术达到了标志性的技术巅峰。

在自然石材结构建筑中，砖常被用于装饰。装饰图案主要靠不同造型的砖块和砖块的各种砌筑形式来实现（图4-5～图4-9）。

图4-8 Muuratsalo summer House, Alvar Aalto, 1952 穆拉萨罗岛实验住宅[1]/夏日别墅，阿尔瓦·阿尔托，1952

在16世纪到17世纪期间，砖块逐步取代自然石材成为建筑承重结构中最主要的材料。

一开始，砖块都是在建筑工地现场制作，直到18世纪开始有专门的制砖厂成立。

图4-9 Yale Laboratory, Philip Johnson, 1963 耶鲁微生物学大楼，菲利普·约翰逊，1963

机器制砖发明于19世纪，这带来了砖结构建筑的新一轮繁荣发展。

---

1 穆拉萨罗岛实验住宅：1952年在芬兰的穆拉萨罗岛，阿尔托为自己设计了一个夏季度假别墅，被称为实验住宅，其最突出的特点是在庭院中用了50多种砖来砌筑铺装墙面地面，具有强烈的实验性质。（译者注）

Railways, the first of which were built in the 1860s, made the transportation of bricks much easier.

The popularity of brick as a loadbearing structure began to subside in the 19th century, when steel structures gained more ground. Finally, brick as a material applied in building frames came to an end at the turn of the 20th century, when reinforced concrete arrived.

The color of bricks varies from creamy white through yellow to red, dark maroon and almost black. The color is determined by the mineral content of the clay, any additional ingredients, as well as the firing method.

A high iron content makes the brick red, while clay rich in kaolin produces light–colored bricks.

By adjusting the firing technique, different surface shades can be achieved; for example, the way the bricks are stacked in the kiln affects how even their coloring is. Prolonged firing results in vitrification and darker colors.

The brick surfaces can be given different finishes: they can be smoothed, brushed to various textures, ripped or moulded to a pattern. The textures are created by shaping the moist clay mass before firing. After firing, the bricks can still be cut or split.

The shape of a brick can be adjusted according to the architect's specifications. Alvar Aalto used bricks of unusual shape in, for example, the façade of the House of Culture in Helsinki.

19 世纪 60 年代火车的出现，使得砖材的运输更为便捷。

19 世纪当钢结构开始获得一席之地时，砖结构作为建筑承重结构的流行趋势逐步消退。最终，20 世纪初，当钢筋混凝土出现后，砖结构不再作为建筑的承重结构。

图 4-10 TKK Main Building, Alvar Aalto, 1964 赫尔辛基理工学院主楼阿尔瓦·阿尔托，1964

砖块的颜色从乳白到黄、红、深栗，甚至黑色都有。砖的颜色取决于黏土的矿物成分和其他辅助原料以及烧制方法。

黏土中铁含量高，烧制的砖为红色，高岭土含量高的黏土产出的砖颜色淡浅。

图 4-11 Fredensborg Housing area, Jorn Utzon, 1963 弗雷登斯堡小区，约翰·伍重，1963

不同的烧制技术，可以形成不同砖块的表面色调。例如，砖块在窑里的放置形式决定了砖块面色的均匀度。延长烧制时间可以使砖面玻化、颜色更深（图 4-10、图 4-11）。

图 4-12 Sydney Opera House, Jorn Utzon, 1973 悉尼歌剧院，约翰·伍重，1973

砖表面可以有各种不同的处理方法：磨平、刷各种纹理、切割或用磨具压制成某一特定形状等。纹理一般是在烧制前在黏土砖坯上刷上去的。烧制完成后，砖块仍可以切割。

根据建筑师的特殊要求，砖块的形状是可以定制的（图 4-12、图 4-13）。例如阿尔瓦·阿尔托在赫尔辛基文化之家的建筑表皮材料就使用了特定形状的砖。

图 4-13 Rautavaara Church, Sakari Aartelo and Esa Piironen, 1982 劳塔瓦拉教堂，萨卡利·阿尔泰洛，埃萨·皮罗宁，1982

The brick conveys the message of earthiness, weightiness and presence. The universal cultural meanings that Reima Pietilä linked with the brick were familiarity, safety and sustainability.

The brick is designed to fit the hand, and is technical in a humane way. In the more upscale buildings, brickwork is often covered with rendering. The projected life of the façade then depends on the durability of the rendering.

Brick symbolizes the symbiosis between the mass and the individual. Industrial production has eradicated the individuality that previous manufacturing methods could give each brick. Each brick was a work of art. An architectural entity was created by thousands of bricks.

The history of brick of humans shows our longing for symbolism towards romanticism.

Maybe this partly explains the wish to create stylistic pastiches in situations where the genuine material is not available or it cannot be used because of its inferior weather performance or other reasons. One such reason is related to status.

At least in Finland, this phenomenon has left an indelible imprint in the formal idiom of wood construction. Even the tradition of using red ochre paint on wooden houses is said to be partly due to the wish to imitate the color scheme of more expensive brick buildings.

Another similar phenomenon has continued in contemporary Finland for at least one generation, as wood–framed houses are given brick cladding. When doing so, fire safety is often one factor, forgetting that fires typically start from the inside.[7]

Brick is one of the oldest standardised building elements. A size that fits the hand, it is easy to lay, even for an amateur bricklayer. Different arrangements and seams can be used to create different surface patterns.

砖块传递着土质、土壤密度、土壤特性等信息。莱玛·比尔蒂拉认为与砖有关的普世文化意义是：亲密、安全和可持续。

砖被设计成适合手大小的尺寸，是一种人性化的技术手段。在比较高档的建筑中，砌砖常常被用作外墙饰面。它的使用寿命取决于粉刷材料的耐久性。

图4-14 Kauniairen Church, Kristian Gullichsen, 1983 考尼艾斯滕教堂，克里斯蒂安·古利克森，1983

图4-15 Poleeni, Pieksämäki, Kristian Gullichsen, 1990 波莱维，皮耶克赛迈基，克里斯蒂安·古利克森，1990

砖是个体与群体共生的象征。工业化生产摧毁了之前手工生产时砖块所具有的独特个性。每一块砖都是一件艺术作品。一个建筑实体是由数以千计的砖块共同创造的（图4-14、图4-15）。

人类社会使用砖的历史表现了人们对于浪漫主义的渴望。这也在一定程度上解释了某一真实材料（砖材）因恶劣天气或者其他原因不能使用时人们创造出仿品的愿望驱动所在。仿品出现的另一个原因是地位。

图4-16 Office Rautio Espoo, Sakari Aartelo and Esa Piironen, 1990 艾斯堡劳蒂奥办公楼，萨卡利·阿尔泰洛，埃萨·皮罗宁，1990

至少在芬兰，这一现象给传统木建筑的形式语言留下了不可磨灭的印迹。用赭石红油漆刷木质建筑表皮的传统很大部分原因是想模仿更为昂贵的砖结构建筑的色彩形式。另一个相似的现象已经在当代芬兰传承了至少一代人：木框架结构的房屋用砖饰面[7]（图4-16、图4-17）。

图4-17 Kauhajoki School of Domestic Economics, Esa Piironen, 1992 考哈约基国内经济学院，埃萨·皮罗宁，1992

砖是最古老的标准化建材之一。砖块大小尺寸称手，易于砌筑，即使对于业余的泥瓦工也很容易。不同的砌筑方式和接缝方式可以创造出不同的墙面形式。

The light and shadows fall at different angles, and a brick surface can therefore have a very vibrant and busy appearance.

Sometimes architects want to create an even brick surface, which can be achieved with paint or a single layer of plaster or rendered to a smooth finish or a course dash coat.

The red hue of the brick sits well in natural environments: the contrast of green plants and trees and red brick has been an inspiration to many architects. Alvar Aalto's architectural red period with red brick is a prime example of this.

Red brick in a white, snow–covered landscape is also very effective.

Bricks come in many colours, depending on the manufacturing technique. Red brick can be fired to achieve many shades from fair ochre to dark maroon.

The yellow brick is commonplace in Continental Europe.

Brick is a fireproof material, and therefore safe.

Ladies and gentlemen, do you know what a brick is? It is a mere trifle that costs eleven cents, a worthless, ordinary object, but it has one unique quality. Give me that brick, and it will immediately become worth its weight in gold.

Frank Lloyd Wright
Milwaukee lecture

不同角度的光影，可以赋予砖块充满活力和富于变化的表面。

有时建筑师需要平滑的砖面，这可以通过刷漆、灰泥饰面、刷平或者浇泼涂层的方式来实现。

红色调的砖在自然环境中非常协调：绿树红墙给了很多建筑师以灵感。阿尔瓦·阿尔托用红砖创造的红色建筑时代是最好的例证。

白色中的红砖——大雪覆盖的红砖建筑也令人印象深刻。

砖块有许多不同的颜色，取决于砖的不同制造技术。通过不同的烧制程度可以获得从赭石色到深栗色多种不同的颜色系的红砖。

黄砖在欧洲大陆随处可见。

砖是耐火建筑材料，因此具有安全性。

你了解砖吗？它是只值区区十一美分一块的廉价常见的东西，但是它有一个独特的品质。给我一块砖，它会立刻变成有黄金般价值的东西。

——弗兰克·劳埃德·赖特
在密尔沃基的演讲

Ceramic tiles

A ceramic tile is a manufactured piece of hard–wearing material such as ceramic, stone, metal, or even glass, generally used for covering roofs, floors, walls, showers, or other objects such as table tops.

The earliest evidence of glazed brick is the discovery of glazed bricks in the Elamite Temple at Chogha Zanbil, dated to the 13th century BC. Glazed and colored bricks were used to make low reliefs in Ancient Mesapotamia, most famously the Ishtar Gate of Babylon (c. 575 BC), now partly reconstructed in Berlin, with sections elsewhere. Mesopotamian craftsmen were imported to the palaces of the Persian Empire such as Persepolis.

Early Islamic mosaics in Iran consists mainly of geometric decorations in mosques and mausoleums, made of glazed brick. Typical turquoise ceramic tiling becomes popular in 10th~11th century and is used mostly for Kufic inscriptions on mosque walls.

One of the best known architectural masterpiece of Iran is the Shah Mosque in Isfahan, from 17th century. Its dome is a prime example of tile mosaic and its winter praying hall houses one of the finest ensembles of cuerda seca tiles in the world.

Ceramic materials for tiles include earthenware, stoneware and porcelain. Terracotta is a traditional material used for roof tiles.

Today ceramic tiles are used in architecture widely, because it is weatherproof and long lasting material on exterior walls.

You can have ceramic tiles in almost every colour. So the best ceramic tile facades are like an art work even today.

瓷砖

瓷砖是用耐磨原料如陶瓷、石头、金属乃至玻璃加工制成的片状材料，一般用在屋顶、底板、墙面、雨篷或者其他物体如桌子等的表面。

迄今发现的琉璃砖最早的使用可以追溯到公元前 13 世纪琼佳藏比（今伊朗胡齐斯坦省境内）的埃兰寺。在古代的美索不达米亚地区，琉璃砖主要用在墙面浮雕装饰上，最著名的是古巴比伦的伊士塔尔城门（公元前 575 年），现今一部分结构被恢复并保存在柏林，另一些则四散在民间。当时美索不达米亚的工匠们被输送到波斯帝国的各大城市如首都波斯波利斯等。

伊朗早期的伊斯兰马赛克用琉璃砖制成，主要由几何图案构成，装饰在清真寺和陵墓里。最典型的绿松石色马赛克在公元 10 世纪到 11 世纪十分流行，常常用来在清真寺墙上篆刻古阿拉伯字母表。

伊朗最著名的建筑杰作之一是始建于 17 世纪的位于伊斯法罕的国王清真寺。它的穹顶是瓷砖建筑的典范，它的冬季祈祷大厅是拥有世界上最好的珐琅彩瓷砖饰面的大厅之一。

制作瓷砖的材料包括陶器、石器和瓷器等。陶土是一种传统的制作屋顶瓷砖的材料。

今天，瓷砖在建筑中被广泛地使用，因为它是一种防水且耐久的外墙材料。

几乎每种颜色的瓷砖你都可以找到。所以，如今精致的瓷砖外墙面就像一件艺术品一样。

Chapter 5　Reinforced as concrete

The purpose of architecture is to bring
the world of material into harmony with
human life.

Alvar Aalto

Concrete is made from a liquid mass which through certain chemical reactions in
its ingredients solidifies into a solid, rock–like consistency.

Concrete is one of the most important and wide–used building materials in the
world. The predecessor of modern concrete, *opus caementicium* was known by
the Romans. The concrete structures built at that time have been preserved until this
day. The reason for the durability of Roman concrete lies in the aluminous volcanic
ash incorporated in it as well as the purity of the limestone. With the fall of the
Roman culture, the use of concrete faded and by the Middle Ages it had become
practically unknown.

The Romans inherited the skill of stone building and developed it further.

They also refined the skill of clay and lime mortar masonry, adopted from
other nations, and combining the two skills resulted in the evolution of
concrete structures.

Lime plaster was introduced in Rome c. 300~250 BC. The new binding agent
pozzolana was discovered by accident c. 200~100 BC, which is when the first
concrete cast walls were built.

第五章　坚固之混凝土

建筑的目的在于使材料
世界与人类生活相和谐。

——阿尔瓦·阿尔托

混凝土是液态物质通过原料间发生一定的化学反应而固化，所形成的像岩石一样的物理性质均匀一致的固体。

混凝土是世界上最重要和使用最广泛的建筑材料之一。现代混凝土的前身是由古罗马人发明的罗马混凝土[1]。古罗马时期的混凝土建筑一直被保护留存到了现在。古罗马混凝土如此耐久的原因在于混凝土中的火山铝灰和高纯度的石灰石成分。随着古罗马文化的消亡，混凝土的使用也随之消退，到中世纪则完全不为人知。

古罗马人巩固了石建筑技术并将其进一步发展。

结合其他国家的经验，古罗马人改良了黏土砂浆砌筑和石灰砂浆砌筑技术，将这两种技术结合，从而促进了混凝土技术的发展。

石膏在公元前300～公元前250年被引入古罗马。公元前200～公元前100年间，当人们建造第一面混凝土现浇墙时，无意中发现火山灰可以作为胶黏剂。

1　混凝土：拉丁语为 opus caementicium。（译者注）

Roman concrete structures were usually built as composite structures with stone and brick. The finest specimen of early concrete building is the Pantheon in Rome, which was built during the reign of Hadrian (118–128 AD).

The brick arches were replaced by concrete ones. The separate elements used in the building supported each other by way of compressive stresses. Continuous cupola structures were introduced. The cupola of the Pantheon with a diameter of 42.2 metres was a world record in its time, broken only in the 20th century, when reinforced concrete was introduced.

A large construction required large scaffolding. Erecting the scaffolding required particular skill.

Concrete building fell into decline with the fall of the Roman Empire. A large notable example of Roman concrete building is the Leaning Tower of Pisa, which was started in 1173.

It was not until the 1700s, when pozzolana regained the interest of scientists and builders. Engineer John Smeatson added pozzolana to the mortar used in the stone foundations of the Eddystone Lighthouse in Plymouth. The lighthouse was completed in 1759. This year is considered to mark the beginning of concrete construction.

Modern–day Portland cement was accidentally invented by the Englishman Isaac Johnson in 1844. This paved way for the growth of the concrete industry.

Cement factories sprang up in Britain, France and later in Germany and Denmark.

Concrete was reintroduced in the 19th century, and in the 20th century, its popularity exploded following the invention of reinforced concrete. The advances made in the casting techniques also increased the potential of concrete. Casting made it possible to use concrete in many shapes and forms.

古罗马的混凝土结构通常是和石材结构、砖结构一起形成混合结构。早期混凝土建筑最好的范本是哈德良统治时期（公元 118~128 年）的罗马万神庙（图 5-1）。

砖拱券被混凝土拱券取代。建筑上各个独立的元素通过元素之间的应力来相互支撑。连续穹顶结构的引入，使得万神庙的穹顶直径达到了 42.2 米，成为当时的世界纪录，直到 20 世纪钢筋混凝土开始使用，这一穹顶直径纪录才被打破。

大跨度的建筑需要大跨度的脚手架。而架设这样的脚手架需要特殊的技术（图 5-2）。

混凝土建筑伴随着古罗马帝国的衰亡而没落。古罗马混凝土建筑中一个著名的案例是始建于 1173 年的比萨斜塔。

直到 18 世纪，火山灰又一次引起了科学家和建造者的兴趣。工程师约翰·斯米顿把火山灰加入砂浆中，用于普利茅斯埃迪斯通灯塔的石基础结构的砌筑上。这座灯塔在 1759 年竣工，这一年被认为是使用混凝土结构的标志元年。

今天的波特兰水泥是在 1844 年由一个叫艾萨克·约翰逊的英国人无意中发明的。这为现代混凝土工业的发展之路奠定了基础。

水泥厂在英国遍地开花，接着是法国，再后来是德国和丹麦。

19 世纪，混凝土被再次使用，到了 20 世纪，钢筋混凝土的发明使混凝土的受欢迎程度大大增加。铸造技术的进步也进一步挖掘了混凝土的潜力，铸造技术使得各种形状和形式的混凝土成为可能（图 5-3）。

In–situ casting was first carried out by packing concrete into wooden moulds without reinforcement. Later, steel reinforcement was added, giving rise to the dominance of reinforced concrete building. The more steel was added, the greater the load structure could withstand.

The final breakthrough for reinforced concrete building was staged at the Paris world fair in 1900 when Francois Hennebique presented his monolithic cast column–and–beam structure, patented in 1892.

The development of the concrete frame enabled free floor plans and non–load–bearing curtain walls. The earliest example of this is Le Corbusier's Dom–Ino concept from 1914.

The base material in concrete is coarse aggregate, making up from 65% to 80% of the total mass of concrete. The properties of the aggregate used, has a great impact on the quality of concrete.

The aggregate can include any sufficiently strong and dense rocks, which do not interfere with the reactions in the cement or compromise the durability of the concrete. The most common aggregates are natural aggregates, which may be naturally produced or mechanically crushed.

Water used in the concrete mix must be pure. The cement reacts chemically with water, causing the concrete to solidify, which makes it possible to cast concrete under water, as long as its is ensured that the cement is not washed away.

Water is also needed in the bonding reaction of the cement but some residual water remains within the concrete. The strength and durability of concrete depends on the water–cement ratio.

现浇混凝土施工最早是将普通混凝土浇到木质模具中。后来，钢筋的加入使得钢筋混凝土建筑迅速崛起。钢筋越多，建筑结构的承重性越好。

钢筋混凝土建筑的最新突破是在 1900 年的巴黎世界博览会上，弗朗索瓦·埃纳比克展示了他的整浇梁柱结构。他在 1892 年申请获取了此项技术的专利。

混凝土框架结构的发展使得自由的平面形式和非承重墙的出现成为可能。最早的例子是 1914 年勒·柯布西耶提出的建筑多米诺框架形式（图 5-4）。

粗骨料是混凝土的基本原料，大约占总配料的 65%～80%。所使用骨料的品质很大程度上影响着混凝土的质量。

骨料可以选用任何一种不会和胶合物发生化学反应或者破坏混凝土耐久性的坚硬密致的石头。最普遍的骨料是自然骨料，直接采自采石场或者开采后经过机器粉碎加工的小石块（图 5-5）。

混凝土中添加的水必须是纯净的水。水泥与水发生化学反应，使得混凝土固化，这使得只要在确保水泥不被冲刷走的前提下，水下浇筑混凝土成为可能。

在水泥的胶结反应中也需要水，但是有一些水分会残留在混凝土中。混凝土的耐久度和硬度取决于水和水泥的配比。

图5-1 Pantheon, Rome, Apollodoros Damascus, 124 AD 万神庙, 罗马, 阿波罗多罗斯 大马士革 公元 124

图 5-2 Colosseum, Rome, 80 AD 大角斗场, 罗马, 公元 8 世纪

图 5-3 Casa Battlo, Barcelona, Antonio Gaudi, 1906 巴特罗公寓, 巴塞罗那, 安东尼·高迪, 1906

图 5-4 Villa Savoye, Le Corbusier, 1931 萨伏伊别墅, 勒柯布西耶, 1931

图 5-5 Fallingwater, Frank Floyd Whrigt, 1939 流水别墅, 弗兰克·劳埃德·赖特, 1939

Cement is a hydraulic bonding agent, which when reacting with water creates a durable end result. The chemical make-up of cement affects the workability of fresh concrete as well as the durability of the hardened concrete. The different properties of concrete, such as the strength, heat development and chemical durability, can be adjusted through the choice of cement.

Cement is made of natural materials: limestone, quartz and clay. Even the Romans mastered the manufacturing of a mass similar to concrete, in which they used the siliceous rock from Mount Vesuvius. The ashy gravel was pulverized and fired at high temperature. The ancient concrete is more durable than contemporary concrete although it has no reinforcements.

Instead of structures cast on site, the norm is nowadays to use factory-made concrete elements. These elements can be column, hollow-core slabs, beams, walls and external wall elements.

The elements can be made to a high degree of finish at the factory; for example, external wall elements come ready with the interior layer, insulation and exterior. Even windows with glass can be added to the element in the factory.

In addition to reinforcement, the surface is one of the key parts of a concrete element in terms of load-bearing and strength, as it usually has to withstand the largest forces.

The surface of concrete elements can be treated with various finishes. It may have an exposed aggregate or polished (mosaic concrete) finish, it can be acid treated or clad with stone, brick or ceramic tiles.

The concrete manufacturing industry is responsible for approximately five percent of all greenhouse gas emissions in the world. Carbon dioxide is generated through the chemical reaction involved in the making of concrete and emissions are also created in the processing of the mass and transportation.

水泥是一种水性黏结剂，它遇水反应并固化，且固化持久不可逆。水泥的化学成分影响着新浇混凝土的工作性能，也影响着混凝土固化后的耐久性。强度、热显影和化学耐久性等性能不同的混凝土可以通过选择不同的水泥来获得。

水泥是使用天然材料制成的——石灰岩、石英石和黏土，它们和古罗马人掌握的混凝土制造技术中的原料非常相似，他们使用维苏威火山的硅质岩。灰色的砂砾通过高温灼烧和压碎，古代的混凝土虽然没有钢筋，却比现代混凝土更为耐久。

现在的标准是采用工厂制作的标准的预制混凝土构件来取代现浇混凝土结构。这些预制构件可以是柱、空心板、梁、墙和外墙构件（图5-6、图5-7）。

这些预制件在工厂中就可以达到很高的完成度。例如，外墙预制件、内壁、保温层都可以在工厂加工，甚至玻璃窗也可以在工厂中安装完成。

除了钢筋外，混凝土表面也是承重和强度的关键部件之一，因为它通常要承受最大的力。

混凝土表面有各种处理方式，包括水洗石表面处理或者水磨石表面处理，也可以做酸化处理或者表面贴石材、砖材、瓷砖（图5-8～图5-10）。

混凝土制造业产生的二氧化碳、甲烷约占全球温室气体排放量5%。二氧化碳在混凝土制造过程中经化学反应产生，大量生产与运输过程中也会出现。

图5-6　Schröder Huis, Gerrit Rietveld, 1924　施罗德赫伊斯住宅，赫里特·里特费尔德，1924

图5-7　Paimio Sanatatorium, Alvar Aalto, 1929　派米奥疗养院，阿尔托，1929

图5-8　TWA Terminal, NYC, Eero Saarinen, 1962　纽约肯尼迪机场TWA候机厅，沙里宁，1962

图5-9　Guggenheim Museum, NYC, Frank Floyd Whright, 1959　纽约古根海姆博物馆，弗兰克·劳埃德·赖特，1959

图5-10　Unite d'habitation, Marseilles, Le Corbusier, 1952　马赛公寓，勒·柯布西耶，1952

The reputation of concrete suffered a serious knock as it became the epitome of the new, pioneering architecture that sought to simplify the use of materials. After Functionalism had stripped all decorativeness from the façades, the next step was to eliminate rendering, paint and other surface treatments that aimed to please the popular tastes.

First came the Neo–Brutalism of the 1950s. To some degree, the awakening of the conservative Britain to architecture happened as a result of that movement from the mid 1950s until the next decade. As the prefabricated housing production simultaneously started to speed up, the triumph of Brutalism extended directly from public building to prefabricated element housing production.[8]

Enter Le Corbusier, who managed to convince us of the virtues of concrete. First was the Unité d'habitation in Marseilles, a new "machine for living". Then he designed the chapel in Ronchamp which finally freed us from the "box" architecture.

At some point, however, the enthusiasm of architects for elements and concrete began to fade. Then it became a bad word, because the quality of concrete architecture reached a low point, beyond acceptable.

In spite of the abstract character of architecture, material has a fundamental effect on the forms produced. Like the other technical components of construction, material is basic to the essence of architecture. It is impossible to make a distinction between the aesthetic and the technical plan. A complete architectural form must be seen as a synthesis of its components.

The surface of concrete varies depending on the material of the mould. The mould is made of either wood, plywood, metal or plastic. Unplanned plank leaves a coarse surface. Splinters from the wood can be felt. Le Corbusier applied this type of concrete in many of his works, which is why he is known as the "father of betón brut". He gained many followers and imitators all around the world in the 1960s.

混凝土因成为一种寻求简洁建筑材料的开创性缩影，从而受到了严重的冲击。在功能主义摈弃了所有建筑表皮的装饰之后，接着就是淘汰抹灰、刷漆和其他用以取悦大众的表皮处理方式。

首先是 20 世纪 50 年代的新粗野主义，从某种程度上讲，是 20 世纪 50 年代开始直至接下来的数十年间保守的英国人在建筑设计上的觉醒改革带来的结果。随着混凝土预制建筑的同时快速发展，粗野主义的成功由公共建筑延伸向预制混凝土住宅产品。[8]

图 5-11　Chapel at Ronchamp, Le Corbusier, 1955　朗香教堂，勒·柯布西耶，1955

接下来是勒·柯布西耶，这个一直致力于向我们展示混凝土优势的建筑师。他设计的马赛公寓是一个全新的"居住机器"。随后，他设计了朗香教堂，让我们最终摆脱了方盒子式的建筑（图 5-11）。

图 5-12　Salk Institute, La Jolla, Louis Kahn, 1963　索尔克研究所，拉霍亚，路易斯·康，1963

然而，在某种程度上，建筑师对混凝土材料的热情开始消退。接着混凝土变成了一个低劣的建筑代名词，因为混凝土建筑的质量达到了一个超出接受范围的质量新低。

图 5-13　Chapel of Holy Cross, Turku, Pekka Pitkänen, 1967　圣十字教堂，土尔库，佩克卡·皮坎嫩，1967

尽管建筑具有抽象性，建筑材料对于建筑形式有着根本的影响。如同建筑的其他技术部件一样，材料是建筑的基础。一个建筑不可能明确区分审美方案部分和技术方案部分。建筑各个部分的完整组合才能被看作一个完整的建筑形式（图 5-12～图 5-15）。

图 5-14　Chapel of Holy Cross, Turku, Pekka Pitkänen, 1967　圣十字教堂，图尔库，佩卡·皮特凯宁，1967

混凝土表面的形式取决于模具的材料。模具一般由木材、胶合板、金属或塑料做成。原木板模具可以形成粗放的表皮，可以让人感受到表皮上残留的木屑。勒·柯布西耶在他的许多建筑作品中都运用了这一建筑表皮形式，让他成为闻名于世的"混凝土之父"。20 世纪 60 年代，他获得了世界各地众多的追随者和模仿者（图 5-16）。

图 5-15　Private House, Kukkapuro, Paloheimo&Ollila, 1969　私人住宅，库卡普罗，帕洛黑莫和奥利拉，1969

More subtle concrete façades were created with plywood or metal moulds.

The quality of the outcome depends heavily on the quality of the concrete used and the skill of the builder. The Salk Institute in La Jolla, California, designed by Louis Kahn, is one of the greatest examples of concrete architecture. Tadao Ando is also one of the great proponents of concrete architecture.

One of architecture's problems is get the form of a construction logically to correspond to the materials, without calling for any kind of causal relationship between material and form, yet decisively rejecting any kind of senseless connection between them. Each material has its own clearly specified technological properties. Though these cannot always be considered the primary criteria for using a material, they nevertheless regulate the ways in which it can be used.

Of the building materials we use today, concrete probably allows the planner the freest hand with form. The concrete mass as such can be cast into almost any mould. Reinforced concrete, the essence of which must always be noted as a combination material supplementing both steel and concrete, of course lays down its own constructional terms. It is processed and formed at one and the same time, on the building site. The raw materials are turned into concrete in the casting moulds in their final and at the same time unique form.

The idea that because of its constructive properties reinforced concrete is better fitted for space structures than for plane structures is rather common. Felix Candela, for example, considers the usual use of steel concrete for frame structures made up of pillars and beams an inheritance from wood and steel structures, and thinks that this works contrary to the nature of reinforced concrete.

It is probably a fact that the bulk of all building will continue to be plane construction. The architectural expression of three dimensional space calls for a composition of the horizontal and the vertical. By the laws of physics and our own equilibrium system, the horizontal plane and the vertical line form the basic three–dimensional system of co–ordination on which all building

我们今天使用的建筑材料中，规划者可以自由地
使用混凝土。混凝土的大量使用也因其几乎可以
用于铸造任何模具。

更精细的混凝土表皮可以通过胶合板或者金属模
具获得。表皮的最终呈现形式主要取决于混凝土
的质量和建造技术（图5-17）。路易斯·康设计的
加利福尼亚州拉霍亚市的索尔克研究中心，是世
界上伟大的混凝土建筑之一。建筑师安藤忠雄也
是混凝土建筑的拥趸之一（图5-18）。

建筑学的一个重要问题是建筑形式与建筑材料的
逻辑对应，无须要求建筑与材料之间的所谓因果
关系，坚决摒弃所谓材料与建筑之间的任何毫无
意义的联系。任何一种材料都有它自身特殊的技
术特性。尽管特性并不完全是考虑使用这一材料
的主要标准，人们还是规范了材料的使用方式。

在今天我们所使用的建筑材料中，混凝土可能是
设计师能够最大限度不受形式限制设计建筑的最
佳材料。混凝土可以浇筑到几乎任何一种模具中。

钢筋混凝土，实质上一直被标注为钢筋和混凝土
互为补充的一种复合材料，当然也有它自身的建
筑结构语言。它是在建筑工地上同时加工与成型
的。原料在模具中最终变成了混凝土，同时也形
成了独特的造型形式。

由于其结构特性，钢筋混凝土结构更适合于空间
结构而非平面结构，这是一个相当普遍的观点。

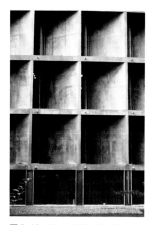

图5-16 Harvard Visual Art Center,
Le Corbusier, 1963 哈佛视觉艺术
中心，勒·柯布西耶，1963

图5-17 Yale Architectural School,
New Haven, Paul Rudolph, 1963
耶鲁建筑学校，纽黑文市，路易斯·
康，1963

图5-18 Koshino House, Tadao
Ando,1981 小筱邸（住宅），安藤
忠雄，1981

is based. Because reinforced concrete is at the moment our most important material for almost all building, it is clear that we cannot restrict its use only to structures whose form corresponds exactly to its constructive character.

Most reinforced concrete is cast into parallelepipedons which do not correspond to the static optimum of the material. However, there forms can be considered sensible and architecturally right, not only in pre–fabricated elements but also in reinforced concrete structures cast on the building site. Basically, the architect's job is to build, to shape form and space by building. Planning must take into account the practical and technical realization of a building. If a construction is to be sensible, its technical realization must also be sensible. Questions of working techniques as well as economics vitally affect the architectural undertaking.

It is characteristic of building in concrete that making the mould usually involves far more work than the actual casting. The problem with concrete structures is largely a problem of the mould. Though the mould only an instrument aiding the realization of a concrete structure, it is even so a vital component. Because the structure takes on its final, fixed shape in the mould, the latter makes its own claims of the structure. The mould is sensible when its form correspond to its material.

It is impossible to force wooden moulds into a shape quite foreign to wood. The final structure produced can only be considered sensible in form when it meets the claims both wood and concrete.

A structure must be considered a synthesis of its own constructive essence and its casting mould. The purely constructive form of a beam is not a parallelepipedon. Even so, it usually takes on its shape as a function of its plank–built mould. Constructive form is thus not always rational. There are, of course, structures in which the optimum constructive form guarantees the optimum results. Many space structures are of this kind, when the structural lightness permitted by the construction usually makes it possible to cover any given span.

Characteristic of rough concrete is complete causality between the mould and the

例如，费力克斯·坎德拉认为继承自木结构和钢结构所通常使用的横梁和柱子组成的混凝土框架结构，是违背钢筋混凝土的自然属性的。

基本的事实是，大部分建筑将继续采用平面结构。建筑所表现的三维空间要求有水平面和垂直面的组合。根据物理学定理和人类的平衡系统，水平面和垂直线条构成了所有建筑以之为基础的三维系统空间。因为钢筋混凝土是当下几乎所有建筑最重要的建筑材料，显然，我们不能仅限于使用与其建构特性完全相符的结构形式。

大多数钢筋混凝土被浇筑成了未能最大限度发挥这一材料静力学特性的方柱体。然而，这种形式，包括预制件和建筑工地上现浇的钢筋混凝土结构，都被认为是合理的和建筑学上正确的。从根本上说，建筑师的工作是建造，通过建造来塑造形式和空间。建筑的设计必须考虑到建筑物的可行性和技术实现。如果一个建筑是合理的，那么它的技术实现也必须是合理的。操作技术问题和经济问题对建筑事业有着至关重要的影响。

混凝土建筑的特点是制造模具的工作量远大于实际浇筑的工作量。混凝土结构的问题主要是模具的问题。虽然模具只是实现混凝土结构的一种辅助工具，但它是一个至关重要的工具。由于混凝土在模具中固化为其最终的形态，模具在结构方面有一定的要求。当模具的形状与混凝土原材料相互契合时，模具便是合理的。

制作一个完全不符合木材特性的木质模具形式是不可能的。只有当混凝土预制件产品的形态同时满足木质模具和混凝土原料的要求，这个预制结构才能被认为是合理的。

预制件必须是综合自身构造性质和浇筑模具的结构。最佳的横梁形式并非方柱体。即便如此，这种形态在木质模具中常被浇筑出来。因此，建筑形式并非总是合理的。当然，在建筑结构中，最优的构造形式是

final structure. Everything which must be in the final structure must appear in the mould, and on the other hand everything in the mould, including every fault, appears inevitably in the final structure. Planning is rather like the work of a graphic artist. The final result must always be thought of as far as possible in the negative.

The primary arrangement of the surface in rough concrete usually stems from the working techniques required and the technological demands of the concrete. The mould material, on the other hand, produces its own surface effect, which is significant mainly as a factor giving metric precision in the structure.

Although the mould structure can be considered a negative, restrictive factor from the point of view of the eventual shape of the concrete, its importance as a factor disciplining the surface may be distinctly positive.

At the moment, planks and plywood board are almost the only mould materials used to cast reinforced concrete on the building site. New mould materials which can be worked in different ways and can be more easily shaped will vitally enrich the forms of concrete architecture.[9]

Concrete is a fireproof material as it is not combustible in itself. It also withstands house fires without collapsing.

Architecture is born of light and material.

Dominique Perrault

最佳建筑效果的保障。许多空间结构都是这样的，当结构很轻巧时，建筑才有可能实现任何给定的跨度。

毛面混凝土的特性完全是由模具和最终结构之间的因果关系造就的。最终结构的每一个细节都必须在模具中有所体现，从另一方面讲，模具上的每一个细节，包括错误，必然体现在最终结构上。设计有点类似平面艺术家的工作。必须要考虑到最终效果可能的消极后果。

毛面混凝土表面最基本的形式设计一般源于操作技术要求和混凝土的技术要求。另一方面，模具材料，作为构造度量精度上的主要因素，对混凝土表面形式也会产生影响。

虽然模具的结构被认为是混凝土塑形的消极制约因素，但模具作为塑造混凝土结构表皮形式的要素来讲无疑具有积极作用。

当下，木板和胶合板几乎是建筑工地浇筑钢筋混凝土的唯一模板材料。新的模板材料必能用各种不同的方式使混凝土造型更简单，最终必将丰富混凝土建筑的形式。[9]

混凝土由于自身不可燃，它是一种防火材料。它能保证着火房屋不发生倒塌。

建筑源自光与材料。

——多米尼克·佩罗

Chapter 6   Thin as steel

Light is the noblest material in
architecture.

Esa Piironen

Steel is a universal term for an alloy of iron and other elements that can be rolled or wrought into shape. The carbon content of steel is 0.03%~1.7%. Alloys with a lower carbon content are called pig iron and those with a higher carbon content cast iron.

I will also be discussing all metals used in construction under the umbrella term of steel: copper, bronze (brass), aluminum, titanium.

The realisation that the green stone in the soil was not, in fact, stone, marked the end of the Stone Age: metals were discovered. With fire, metal could be extracted from the ore and, as it was malleable, it could be worked into different shapes. Metal tools developed.

Metals are made up of crystals that can be reformed without cleaving them: the atoms only rearrange themselves. Metals are structurally harder than wood or stone.

When the "green stone", which in reality was copper, and tin were mixed, the result was bronze and with that discovery began the Bronze Age.

C. 1200 BC was the beginning of the Iron Age. Arms and shields evolved.

Although steel was already known in China in 200 AD, it was not until 1855,

第六章　纤巧之钢

光是建筑的最佳材料。

——埃萨·皮罗宁

钢材是铁合金和其他可以轧制或加工成形的金属物质的一个通用术语。它的碳含量为 0.03%～1.7%。碳含量较低的铁合金叫生铁，碳含量高的称为铸铁。

这一章中，我们会讨论在建筑结构中用到的所有统称为钢材的金属材料，包括铜、青铜（黄铜）、铝、钛等。

发现土壤中的绿石实际上并非岩石，这标志着石材时代的终结：金属被发现了。金属可以用火从矿石中提取出来，因为可铸造性，金属可以被加工成各种形状，从而发展出了金属工具。金属由晶体组成，晶体可以重新组合而不被破坏，即仅原子顺序重新排序。金属在结构上比木材和石材更加坚硬。

当"绿石"——事实上是铜、锡混合矿石——最终带来青铜的发现，青铜器时代到来了。

公元前 1200 年是铁器时代的开端。武器和盔甲得到了发展。

虽然在公元 200 年的中国就有了钢材，但是直到 1855 年，贝氏炼钢炉

when the Bessemer converter was patented, that the cost of producing steel came to match that of cast iron. In the early days of industrial revolution, steel had been too expensive for generic use and structures, even ships, were mainly built from cast iron. The Bessemer process enabled manufacturing at a much lower cost, and nearly all manufacturing became steel–based.

Historically, steel was first made with a method similar to the Bessemer process in China during the Qin dynasty. Air was blown into molten iron, burning off the excess carbon.

The method of crucible steel was discovered in India in the 4th century AD. In this method, rod iron was heated in a crucible with crushed glass and carbon. As the glass melted, it absorbed the impurities in the iron and the carbon was bound by the solid iron. The skill of making crucible steel spread to Europe but was forgotten with the fall of the Western Roman Empire.

The Middle Ages saw the development of alloy steel. In this method, cast iron and rod iron is forged together to create wrought steel. The smiths of Damascus developed their steel–making skills to an extreme, resulting in the invention of Damascus steel.

In the Viking Age, swords made from this metal were world–famous, as they were both hard and ductile.

It was not until the 18th century, however, that metals were introduced to the construction industry.

When the technology improved enough to allow for the removal of impurities from iron and the technique of tempering was developed, the manufacture of steel began. in 1863, Henry Sorby discovered through his microscopic studies that to achieve high–quality steel, the correct amount of impurities was in fact necessary, as was added carbon.

Henry Bessemer developed a converter which enabled the industrial production of steel. First, all carbon was removed, and then 1 per cent carbon added, a

申请到专利，才使生产钢材的成本和铸铁的成本接近。工业革命的早期，钢材太贵，无法在平时使用，也不能用在建筑结构上，甚至造船，用的也是铸铁。贝氏炉炼钢法使钢材制造成本大幅降低，使得几乎所有制造业都以钢材为基础。

根据历史记录，钢材最早是在中国秦朝用一种类似于贝氏炉炼钢法炼制得到的。空气吹入铁水中，燃烧掉多余的碳。

公元前 4 世纪，印度发明了坩埚炼钢法。在这一方法中，条状铁、碎玻璃和碳被放在坩埚中加热。当玻璃熔化后，会吸收铁中的杂质，而碳就最终结合到铁中一起固化。这一技术被传到了欧洲，但是却随着古罗马帝国的消亡而被遗忘了。

中世纪见证了合金钢的发展。在这种方法中，铸铁和棒铁一起被锻造成锻钢。大马士革的铁匠们把这一炼钢技术发展到了极致，最终发明了大马士革钢。

北欧海盗时代[1]，用大马士革钢铸的剑，由于其坚硬而有韧性，闻名于世。

直到 18 世纪，金属才被引入到建筑业。

当技术改进到足以去除铁中的杂质时，回火技术得到发展，钢材制造业开始了。1863 年，亨利·索比通过显微镜研究发现：事实上，适量的杂质添加，如碳的添加，对于得到高质量的钢是必要的。

亨利·贝塞姆开发的转炉使得工业生产钢材成为可能。第一步，所有的碳被清除，然后添加 1% 的碳，这归功于罗伯特·马希特的一

---

1 北欧海盗时代（Viking Age，8 世纪末～11 世纪），斯堪的纳维亚海盗对欧洲国家进行海上贸易与抢劫商船活动的时期。（译者注）

discovery credited to Robert Mushet.

Metals are malleable materials. Metals are weakened by fatigue and yield under stress.

Besides carbon, steel is alloyed with many other elements to achieve the desired properties, including nickel, silica, chromium, cobalt, tungsten, molybdenum, vanadium and aluminum.

Steel also plays a key role in reinforced concrete structures. Similarly, steel structures have been essential in the construction of skyscrapers from the very beginning. Even today, steel is the main frame material in high–rise buildings.

Man has always tried to substantiate the meaning of his existence by erecting public buildings at the limits of his technical abilities. The value of the building materials used thereby was redefined again and again over time.

However, passed has indeed the architectural phase which emphasized shapes at the expense of materials. Today, the question for the meaning of different materials has come back to being relevant.

Architect Louis Kahn asked the brick what it would like to be. Steel might be asked the same question in various ways. Back in his days, Buckminster Fuller questioned how much a building weighs. Australian Glenn Murcutt in turn shares the lesson he learnt from the aborigines: we should ensure not to touch this planet too much.

The awareness for environmental restrictions has also introduced new perspectives to architecture. What is important in steel construction is to know the material's basic properties. Steel is more resistant to traction than to compression–which constitutes the base for its expressive potential.

Around the mid–19th century, after the turn of the century and then again in

个发现。

金属材料可塑性强。金属在重锤下变弱，在压力下变形。

除碳外，钢还可以与许多其他元素合金化，以达到想要的性能，包括加入硅、镍、铬、钴、钨、钼、钒和铝等。

钢在钢筋混凝土结构中也起到关键作用。同样，从一开始，钢结构在摩天大楼的建设中也是必不可少的。时至今日钢材仍是高层建筑的主要结构材料。

人类总是试图通过挑战自身技术能力极限而建造出公共建筑以证明其存在的意义。随着时间的推移，所使用的建筑材料的价值被一次又一次地重新定义。

然而，以材料为代价的强调形式的建筑时期已经过去了。现在，表现不同材料的意义变得更有价值。

建筑师路易斯·康问一块砖想成为什么，钢也可以以不同的方式问同样的问题。回到巴克敏斯特·富勒的时代，他问建筑的重量。澳大利亚人格伦·默科特紧接着分享了他从土著人那里学到的经验：我们应轻轻触碰这个星球。

环保意识的觉醒也给建筑带来了新的视角。在钢结构中，最重要的是要知道材料的基本性能。钢材的抗拉性优于抗压性——这是挖掘钢材潜能的基础。

大约在 19 世纪中期，世纪之交之后和 20 世纪 60 年代，钢铁产量都有

the 1960s, the production of steel experienced a significant increase. These milestones laid the foundation for contemporary steel construction which currently seems to be particularly en vogue. The difficulties in steel construction are no longer of technical but of cultural nature–architecture relies on more than bare material–it also requires mind.

By the end of the 18th century, iron was the material mainly used for primary structures. Initially, supporting structures were made from cast iron, later from wrought iron. In the mid–1850s, steel production was commenced, and already by the end of the century steel had become the material then mainly used for supporting structures.

Bridges were the first large–scale iron buildings. The world's first iron bridges panning across River Severn in Coalbrookerdale, England–was built from 1776 to 1779. The building material for this bridge with its clear span of 30 meters was cast iron.

About the same time, first experiments with using cast iron in building construction took place in England, where cotton industry and its machinery required increasingly large premises. Timber columns and beams were replaced by iron structures. The most famous structure of that time was built by Boulton and Watt for Philip&Lee in Lancashire. The respective building had seven stories, and its intermediate floors were supported by a kind of steel beam. The steel beams' fire safety and load capacity clearly exceeded those of timber structures.

Ever since the early 19th century, steel has also been used for building domes and vaults. The iron dome of Halle au Blén in Paris (1808) was the world's first dome completely made from steel.

大幅度的增长。这些里程碑为现在看来似乎特别流行的现代钢结构奠定了基础。钢结构的难点不再是技术上的而是文化性上——建筑不再仅仅依赖于材料，它还需要精神。

18 世纪末，铁仅用于基本结构。起初，支撑结构用的是铸铁，后来用锻造铁。19 世纪中叶，钢铁开始生产，到 19 世纪末，钢材成为主要的支撑结构材料（图 6-1）。

第一座大型的钢铁建筑是桥梁。世界上第一座铁桥为英国的科尔布鲁克代尔大桥，横跨塞文河，建于 1776~1779 年间。这座净跨度 30 米的桥梁，使用的建筑材料是铸铁（图 6-2）。

在同一时期，英国的棉纺业及其机械化生产需要越来越大的厂房，从而产生了在建筑施工中使用铸铁结构的首次试验。木质梁柱被钢结构取代，当时最著名的铸铁结构建筑是由博尔顿和瓦特在兰开夏郡[1]为菲利普 & 李公司建造的。这一建筑有七层，每层均由钢梁支撑，钢梁的防火性能和承重明显地超越了木梁结构。

自 19 世纪初以来，钢也用来建造建筑穹顶和拱顶。1808 年巴黎的布伦集市大厅的铁穹顶是世界上第一个用钢材建造的穹顶。

图 6-1　Tower Bridge, London, 1894　伦敦桥的钢结构基础塔楼，1894

图 6-2　Coalbrookedale bridge, 1779　科尔布鲁克代尔大桥，1779

1　兰开夏郡（Lancashire）是英国英格兰西北部的郡，包含 2 个城市，是英国工业革命的发源地。16~18 世纪亚麻、毛、棉纺织工业迅速发展，成为全国最大的纺织工业区。（译者注）

At the beginning of the century, numerous fairground halls with steel and glass roofing were built in Central Europe. These galleries still make a majority of the roofed public spaces in many European city centers. The most famous amongst them are Galerie d'Orleans (1829) in Paris, Galeries St. Hubert (1847) in Brussels, Vittorio Emannuele II (1867) in Milan and GUM department store (1893) in Moscow.

Closely related to steel construction also is the technical development of the rail system–steel as material enabled wider ranges. The following train stations in London and Paris have for instance been roofed with respective steel structures: Gare du Nord (1847), Gare de l'Est (1852), Paddington Station (1847) and King's Cross Station (1852). They were followed by numerous other stations, roofed with glass and steel in the same manner.

Among the most interesting developments in architecture is the decision to use steel and glass also for the construction of greenhouses. Kew Gardens, the royal botanic garden near London, provides such a palm house built in 1848, which served as model for numerous later–built glasshouses.

Several world exhibitions were occasion for exploring the limits in steel construction. Without a doubt, the respective peak was reached 1851 in London: Joseph Paxton's Crystal Palace still acts as model for contemporary steel constructions. Copies of this building can be found all over the world. The Crystal Palace's key role in the history of architecture is mainly based on its size, the speed of its erection, the involved standardization and industrial assembly. Already back then, this building met today's requirements in terms of recyclability and versatility. At the end of the fair, the building was taken down and subsequently re–erected in Sydenham. However, the Palace was later destroyed by fire.

19世纪初，在中欧建造了无数屋顶为钢架和玻璃结构的集市大厅。这些长廊大多数依然占据着许多欧洲城市中心的室内公共空间。其中最著名的是巴黎老王宫奥尔良廊（1829），布鲁塞尔的圣·休伯特购物长廊（1847），米兰维托里奥·埃马努埃莱二世购物长廊（1867），莫斯科古姆百货中心（1893）（图6-3）。

图6-3 GUM, Moscow, Alexander Pomerantsev, 1893 古姆百货，莫斯科，亚历山大·波梅兰采夫，1893

与钢结构密切相关的是铁路系统的技术发展，钢材在更广泛的领域被使用。例如，在伦敦和巴黎的下列火车站都有钢结构屋顶：火车北站（1847），巴黎东站（1852），帕丁顿火车站（1847），国王十字车站（1852）。之后还有无数车站都采用了同样的玻璃和钢架屋顶。

建筑中最有趣的发展之一是用钢和玻璃来建造温室。伦敦附近的英国皇家植物园，建造了一个棕榈植物的温室，成为之后众温室的典范。

多次世界博览会都为探索扩展钢结构的使用范畴提供了极好的机会。毋庸置疑，探索的巅峰之作是1851年由约瑟夫·帕克斯顿设计的伦敦水晶宫（图6-4），时至今日仍然是当代钢结构的典范之作。在世界各地都能找到这个建筑的各种翻版。水晶宫在世界建筑史上的重要地位主要基于其规模、安装速度、相关的标准和工业化装配。放在现在，这个建筑也依然符合现今建筑在循环利用和多用途方面的要求。在博览会结束后，水晶宫被拆除，随后在西德纳姆重建。然而，不久被大火所毁。

图6-4 Crystal Palace, London, Joseph Paxton, 1851 水晶宫，伦敦，约瑟夫·帕克斯顿，1851

For the 1889 Paris International Exposition, a number of buildings of leading roles in the history of steel construction were erected. Without a doubt the most famous of them is Eiffel Tower. Gustav Eiffel's extensive career as bridge builder came to its peak by this construction, which later became landmark of Paris.

The tower was clearly disliked by many of his contemporaries. But nevertheless, it verifies actual technological advantage: the air space contained by the Eiffel Tower's structures weighs more than the actual steel structure does.

Also the machine hall (Galerie des Machines), designed by architect Ferdinand Dutert and erected on the occasion of the Paris International Exposition, was demolished later. Only Grand and Petite Palais still exist today to report of this period in steel construction.

In the late 19th century, steel was furthermore the most popular building material for department stores. Until the present day, Paris malls such as Bon Marché (architect Louis–Charles Boileau) built 1876, Le Printemps (architect Paul Sedile) built 1885, La Samaritaine (architect Franz Jourdain) built 1910 and Les Galeries Lafayette (architect Ferninand Chanut) built in 1912 inspire new department store buildings to the very detail.

The world's first higher steel frame buildings were erected in the 1880s in Chicago. The city's rapid growth had boosted real estate prices, and people were forced to develop the city into a new direction: upward. This is how skyscrapers came into being. 14–floor Tacoma

1889 年的世界博览会，矗立起许多在钢结构建筑历史上有主导意义的建筑。毫无疑问，埃菲尔铁塔是最著名的。因为这座塔，古斯塔夫·埃菲尔作为桥梁建筑师的职业生涯达到了巅峰，这座塔最终成为了巴黎的地标性建筑（图 6-5）。

显然，他同时代的很多人都不喜欢这座塔。尽管如此，它证实了客观存在的技术优势：埃菲尔铁塔容纳的空间内能承受重量比铁塔结构自身更重的重量。

建筑师费迪南德·杜特设计的矗立在巴黎世博会的主馆机械馆，后来也被拆除，如今只剩下大宫和小宫留存，向人们展示着当时的钢结构。

19 世纪末，钢材成为百货公司建筑最受欢迎的建筑材料。直到今天，巴黎的许多购物中心，如建于 1876 年的乐蓬马歇百货（建筑师路易斯·查尔斯·布瓦洛）、1885 年建的巴黎春天（建筑师保罗·塞德勒）、1910 年建的莎玛丽丹百货公司（建筑师弗兰兹·约旦）、1912 年建的老佛爷百货（建筑师费尔南德·沙努特）等，都为当今新建的百货大楼建筑带来许多细节上的灵感（图 6-6）。

19 世纪 80 年代，芝加哥建成了世界上第一个高层钢结构建筑。城市的迅速发展推动了地产价格上涨，人们被迫向新的方向发展城市：向上。摩天大楼就是这样诞生的。1884 年，14 层的塔科马大厦竣工。大厦的入口支撑结构使用了钢梁柱——与以往不同，外墙不再是主要结构。电梯的投入

图 6-5　Eiffel Tower, Gustaf Eiffel, 1889　埃菲尔铁塔，古斯塔夫·埃菲尔，1889

图 6-6　Galeria Vittorio Emannuele Ⅱ, Milan, Giuseppe Mengini, 1867　维托里奥·埃马努埃莱二世购物长廊，米兰，1867

Building was completed in 1884. The building's entire support structure was made from steel columns and beams–in contrast to what was common before, its outer walls served no longer as primary structures. This new type of building was facilitated by the availability of electric elevators(Otis). Starting from Chicago, high–rise buildings first conquered the rest of the United States and then Europe. By the end of the century, George Wyman's Bradbury Building in Los Angeles from 1893 was the most detailed and technologically advanced steel construction ever seen. The technology's ongoing advancement led to increasingly higher buildings. The Empire State Building erected in 1931 was the world's first building higher than the Eiffel Tower. It remained the world's highest building until New York's World Trade Center was completed in 1973. One year later 445 meters high Sears Tower was built in Chicago. It wasn't topped before the Petronas Towers were built in 1997 (architect Cesar Pelli) in Kuala Lumpur. After that the highest building was in Taipei for some time before they built even higher Burj Khalifa in Dubai.

The European early 1900s years were greatly influenced by Jugendstil. Natural shapes were also emphasized in structural steel works.

Ever since the turn of the century, Hector Guimard's steel constructions for the Parisian Metro have been vital part of the setting. Another example for his time is Belgian architect Victor Horta, whose works combine steel structures, natural stone and ceramic tiles at high quality. Jugendstil architecture (art nouveau) preceded structuralism. After World War I, it quickly forfeited its avant–garde position.

In the US, industrial production developed quickly as a consequence of applied technology. Henry Ford's automobile production greatly influenced human mobility.

Steel was increasingly used as primary material in the construction of factory buildings. Albert Kahn designed the first plants for Ford and later on factories in

使用促进了这一新建筑类型的发展。从芝加哥开始，高层建筑征服了美国其他地区，继而征服了欧洲。到 19 世纪末，1893 年乔治·怀曼设计的洛杉矶布拉德伯里大楼是有史以来最细致和最先进的钢结构建筑（图 6-7）。钢结构技术的不断发展带来了高层建筑的迅速发展（图 6-8）。1931 年建造的帝国大厦是世界上第一座比埃菲尔铁塔高的建筑大楼（图 6-9）。直到 1973 年纽约世贸大厦建成，帝国大厦一直保持着世界最高建筑的记录。一年后，445 米高的西尔斯大厦在芝加哥竣工，这是 1997 年吉隆坡双子塔（建筑师塞萨尔·佩利）建成前的世界第一高楼。之后，最高的建筑出现在中国台北，不久后更高的哈利法塔在迪拜建成。

20 世纪早期的欧洲深受新艺术运动的影响。钢结构工程中也强调自然的造型。

从 20 世纪初开始，赫克托·吉马尔所设计的巴黎地铁钢结构成为这一运动的重要组成部分。同一时期的另一个著名代表是比利时建筑师维克多·霍塔，他的作品高质量地将钢结构、自然石材和瓷砖结合在一起。新艺术风格建筑之后是结构主义建筑。第一次世界大战后，新艺术风格迅速丧失了其先锋地位。

在美国，应用技术的发展带来了工业生产的迅速发展。亨利·福特的汽车生产极大地影响了人类的流动性。

钢材迅速成为厂房建筑的主要结构材料。阿尔伯

图 6-7　Bradbury Building, LA, George Wyman, 1893　布拉德伯里大楼，加州洛杉矶，乔治·怀曼，1893

图 6-8　Chyrler Bldg NYC, William van Alen, 1931　纽约克莱斯勒大厦，威廉·凡·艾伦，1931

图 6-9　Empire State Building, NYC, Shreve, Lamb and Harmon, 1931　帝国大厦，纽约，施里夫，兰博＆哈蒙建筑公司，1931

the Soviet Union. His rational design had lasting effects on the emergent functional architecture.

At the beginning of the century, contemporary architecture gradually began to gain acceptance. Its development was not only influenced by technical innovation but also by desire to design a new use of forms.

Again steel as building material played a major role. The factory halls built in 1910 by architect Walter Gropius carried on the style already previously popular in the US. In 1926, the Dessau Bauhaus school proclaimed the dawn of a new era.

Le Corbusier, regarded a modernism pioneer, used mainly concrete for his buildings. Nevertheless, his concepts of functionalism were revolutionary. Le Corbusier deployed steel in few buildings only. The most famous of them are Pavillon Suisse in Paris (where steel served as frames for facades) while concrete was used for structure and the Small Exibition Hall for Heidi Weber in Zurich.

Mies van der Rohe, another trailblazer of the time, later got famous for paving the way for steel architecture, and his influence is still obvious in each and every steel construction. The Barcelona Pavilion built in 1929 produced new approaches to arranging steel details. Mies van der Rohe's dictum "God is in the details" perfectly describes his architectural philosophy. After immigrating to the United States, he designed a number of buildings that left their marks on the history of architecture.

Best known among them are the premises of the Illinois Institute of Technology in Chicago (1951~1956), Seagram

特·卡恩为福特设计了第一批工厂，后来又在苏联设计了工厂。他的理性设计对新兴的功能性建筑产生了持久的影响。

20世纪初，当代建筑逐步获得认可。当代建筑的发展不仅是因为受到技术革新的影响，而且也是对新的设计形式的期待。钢材再一次成为最重要的建筑材料（图6-10）。

1910年建筑师瓦尔特·格罗皮乌斯设计的工厂建筑所开创的风格早已风靡美国。1926年，德骚的包豪斯新校舍建筑真正带来了新时代建筑的曙光。

勒·柯布西耶，公认的现代主义先锋人物，他的建筑以混凝土材料为主。然而，他的功能主义概念极具革命性。勒·柯布西耶仅在少数几个建筑中用了钢结构。其中最著名的是巴黎国际大学城的瑞士学生会馆（钢架结构，混凝土饰面的一个建筑）和瑞士苏黎世的海蒂·韦伯中心（图6-11）。

密斯·凡德罗，同时代的另一位现代主义建筑先驱，作为钢结构的开路先锋而闻名于世，他的影响力依旧渗透于每一个钢结构建筑。1929年巴塞罗那博览会德国馆采用了新的钢构架细部节点的布置方法。

密斯·凡·德罗的格言"细节就是上帝"完美地阐述了他的建筑哲学。在移民美国后，他设计了许多名垂建筑史的建筑。最著名的是芝加哥伊利诺伊理

图 6-10 Sears Tower, Chigaco, SOM, 1937 西尔斯大厦，芝加哥，史欧姆公司设计，1937

图 6-11 Heidi Weber Gallery, Zurich, Le Corbusier, 1963 海蒂·韦伯博物馆，苏黎世，勒·柯布西耶，1963

Building in New York (1957) and the Berlin National Gallery in (1968).

His architecture relies on realistic structures and extensive details. It was not before post–modernism that Mies' dictum "less is more" was superimposed by a new concept. With his book "Complexity and Contradiction in Architecture" (1966), Robert Venturi released architecture. According to him "less is a bore" (instead of being more).

Built in 1931 in Paris, La Maison de Verre (architects Pierre Chareau and Bernand Bijvoet) is one of the gems in steel construction. The building's richness in detail and glass tiles combined with its steel framework make it really unique. The house that Ray and Charles Eames designed in Santa Monica, USA (1949) for themselves inspired in turn prefab house architecture all over the world.

Another great name in steel construction is Jean Prouvé. He was one of those architects seeking inspiration in the automobile, aviation and shipbuilding industry, which he reflected in his steel architecture design vocabulary. The clear lines in his construction are closely linked to the industrial style of that time.

One of the key influences in this century is Buckminster Fuller—the American innovator and designer still shapes contemporary steel architecture. His geodesic domes may not have gained the attention he aspired but his concepts on energy saving, lightness and airy structures are highly topical even to the present day. His most famous vision was roofing the center of Manhattan by a screen made from steel and glass which has, however, not been realized so far.

工学院（1951～1956）、纽约西格拉姆大厦（1958）（图6-12）和柏林新国家美术馆（1968）。他的建筑主要有赖于可实现的结构和大量的细节设计（图6-13）。

后现代主义的到来，密斯的格言"少即是多"被叠加了新的涵义。1966年罗伯特·文丘里在其著作《建筑的矛盾性和复杂性》一书中重新阐释建筑，根据他的理论，"少就是乏味"（而非多）。

1931年建于巴黎的"玻璃之家"（建筑师皮埃尔·夏罗和伯纳德·毕吉伯）是钢结构的稀世之作。这一建筑细节丰富，玻璃砖和钢框架使得这个建筑相当独特。1949年雷与查尔斯·伊姆斯在美国加州圣莫妮卡为他们自己设计的别墅，给全世界预制房屋建筑带来了灵感（图6-14）。

钢结构建筑史上另一个重要人物是简·普鲁威（Jean Prouvé）。他是从汽车、航空和船舶工业中寻找灵感的设计师之一，他将这些灵感反映在他的钢结构建筑设计语汇中。他建筑中清晰的线条和当时的工业风格有着密切的联系。

20世纪还有一位具有重要影响的人物是巴克敏斯特·富勒——美国发明家和设计师，他仍然影响着当代的钢结构建筑。他的圆屋顶测量法可能并没有让他获得他自己所期待的关注，但是他提出的节能、轻盈、通风结构的概念在今天依然是非常热门的话题。他最著名的构想是用钢架和玻璃做成一个巨型穹顶，将曼哈顿中心覆盖起来，然而这一构想至今未能实现。

图6-12 Seagram NYC, Mies van der Rohe, 1958 纽约西格拉姆大厦，密斯·凡·德罗，1958

图6-13 Farnsworth House, Plano, Ludwig Mies van der Rohe, 1951 范斯沃斯住宅，普莱诺，路德维希·密斯·凡·德罗，1951

图6-14 House #8, Santa Monica, Ray and Charles Eames, 1949 八号住宅，圣莫妮卡，雷与查尔斯·伊姆斯，1949

These days, steel architecture increasingly tends towards high–technology. So–called high–tech architecture frequently simulates acrobatics and is influenced for instance by aerospace technologies.

Today's most influential steel construction architects are Europeans. Mainly England and France compete for the leading role in steel construction.

Centre Pompidou in Paris marks the beginning of contemporary steel architecture. Meanwhile, its creators Richard Rogers and Renzo Piano go separate ways but are still leading in steel construction. Other famous British steel architects are Norman Foster, Nicholas Grimshaw, Michael Hopkins and Ian Ritchie. Structural designers Peter Rice and Ove Arup were also important structural engineering constructors.

In France, steel architecture development is driven by Jean Nouvel, Dominique Perrault, Paul Andreu and Odile Decq.

In turn, the Americans I.M.Pei, Richard Meier and Frank Gehry designed a number of public buildings in Europe that feature diverse forms of steel as major building material. Furthermore, steel construction is booming in Japan where various European architects design public buildings made from steel.

Being inspired by Mies van der Rohe, rationalistic architecture was most popular in Finnish concrete and steel construction. Even Alvar Aalto was enthusiastic about this style, and his design works for the façade of "Rautatalo" building (1954) have been directly inspired thereby.

如今，钢结构朝着高科技方向迅速发展。所谓的高技术建筑常看似悬浮飞行，受到航天航空技术的影响。

今天最具影响力的钢结构建筑师都是欧洲人。主要是英国人和法国人在争夺钢结构的领导地位。

巴黎蓬皮杜艺术中心标志着当代钢结构建筑的开端（图6-15）。它的设计师理查德·罗杰斯和伦佐·皮亚诺虽然分道扬镳，但依然引领钢结构潮流（图6-16）。著名的英国钢结构建筑师有诺曼·福斯特、尼古拉斯·格里姆肖、迈克·霍普金斯和伊恩·里奇。结构设计师彼得·莱斯和奥·艾拉普也是重要的钢结构工程设计师（图6-17、图6-18）。

在法国，推动钢结构发展的建筑师主要有吉恩·诺弗尔、多米尼克·佩罗、保罗·安德鲁和奥迪尔·德克（图6-19）。

美国建筑师贝聿铭、理查德·迈耶和弗兰克·盖里在欧洲设计了许多用不同形式的钢结构作为主要建筑材料的公共建筑。与此同时，多位欧洲建筑师在日本设计了钢结构公共建筑，使得钢结构建筑在日本蓬勃发展（图6-20）。

受密斯·凡·德罗的启发，混凝土和钢结构的理性主义建筑在芬兰非常受欢迎。即使是阿尔瓦·阿尔托也热衷于这种风格，他设计的拉塔塔罗商业中心大楼（1954）的外立面就是受这一风格直接影响的作品。

图6-15 Pompidou Center, Paris, Renzo Piano and Richard Rogers, 1977 蓬皮杜中心，巴黎，伦佐·皮亚诺、理查德·罗杰斯，1977

图6-16 Lloyd's, London, Richard Rogers, 1986 劳埃德大厦，理查德·罗杰斯，1986

图6-17 HSBC, HongKong, Norman Foster, 1986 汇丰银行总部大楼，诺曼·福斯特，1986

图6-18 Waterloo Terminal, London, Nicholas Grimshaw, 1993 滑铁卢车站，伦敦，尼古拉斯·格里姆肖，1993

In the 1960s, some smaller buildings–such as "Allas Matilainen" in Turku in 1967 (designed by architecture students Esa Piironen and Mikko Pulkkinen), the Hyrylä Parish Center in 1967 (architects Kirmo Mikkola and Juhani Pallasmaa) and the extension of Yläne Dancing Hall in 1968 (architect Ola Laiho) were erected in adherence to this style. At the same time, Finland's largest suspension bridges were completed in Sääksmäki (1963) and Kirjalansalmi, Parainen (1964).

In consequences of numerous factors, architecture grew increasingly varied and richer throughout the centuries. In that progress, the social and ethical climate played just as much a role as prosperous economic trends (especially after World War II) and advancements in technology and building material did. By the end of the century, steel in public construction had undergone numerous changes. As before, its advantages are still about speed and light weight, about its diversity, dimensional accuracy and variability.

From an ecological point of view, we should use as little of this material as possible in order to save natural resources. Steel buildings should be thin and light–weight–especially because this material is not a regenerative one.

Recyclability is one of the major aspects in the building industry's sustainable development. Steel, for instance, is fully recyclable. Even Paxton's Crystal Palace was relocated without any problems. By now, almost 80% of all steel used is already made from recycled material.

20 世纪 60 年代，一些较小型的建筑，如 1967 年建筑学学生埃萨·皮罗宁和米科·普尔金嫩设计的建于芬兰图尔库的"Allas Matilainen"，1967 年建筑师克拉默·米科拉和尤哈尼·帕拉斯玛设计的 Hyrylä[1] 行政中心，以及 1968 年建筑师奥拉·拉霍设计的伊兰舞蹈厅扩建工程都是在追随这个风格的基础上建起来的。与此同时，芬兰最大的悬索桥于 1963 年在 Sääksmäki（芬兰的一个小镇）竣工，1964 年位于帕莱恩（芬兰西南部的一个镇）的悬索桥竣工。

图 6-19　National Library, Paris, Dominique Perrault, 1995　国家图书馆，巴黎，多米尼克·佩罗，1995

在众多因素的影响下，近百年来建筑发展得更加丰富与多样化。在这一发展过程中，社会和伦理氛围在建筑材料和技术的进步中发挥着和繁荣的经济趋势同等重要的作用（尤其是第二次世界大战以后）。20 世纪末，公共建筑中的钢结构技术经历了无数次变革。和以往一样，钢结构的优势依然在于安装速度和质量轻上，在于它的结构多样、尺度精确和可变化性。从生态学的观点看，我们应该尽可能地少用钢材来保护自然资源。因为这种材料的不可再生性，钢结构建筑应尽可能小体量和轻巧。

图 6-20　Guggenheim Bilbao, Frank Gehry, 1997　古根海姆博物馆，毕尔巴鄂·弗兰克·盖里，1997

可回收再利用是建筑业可持续发展的主要考量因素。例如，钢材是可充分回收利用的。甚至水晶宫也曾毫无问题地易地再建。到目前为止，几乎 80% 的钢材是回收再利用得来的。

1　Hyrylä：图苏拉市的三镇之一，行政中心，大约十分钟车程到赫尔辛基万塔机场，距赫尔辛基 25 千米左右。（译者注）

The combination of glass and steel yielded some of the most beautiful architectural works of our time. Every once in a while, the glass's transparency and the steel's lightness even result in an illusive sight, reflecting modern expectations towards democracy: public sphere and clarity.

Another challenge in architecture lies in overcoming gravity–and steel provides ideal conditions for this.[10]

When designing steel buildings, I have tried to use as less material as possible. As a young architectural student I admired Mies van der Rohe's architecture and details. Details were very important at that time.

Later I found that the idea of architecture is not only details, but the comprehensive design solution, that can heal inhabitant's environment with all aspects of architecture. So the big issues became more important.

And because the building industry could not come to the same quality level as automobile industry, air plane industry and not even ship building industry, I concentrated more to healing architecture.

I started to use more steel in my design work. It is a modern material and when it is also a renewable material through recycling, it became my favorite material designing railway stations and subway stations.

Best example of my design work is Helsinki Railway Station platform roofing, which was based on an international architectural competition. For me to win Moneo, Foster and other famous architects in a competition, was a great experience.

钢架和玻璃的结合产生了不少我们这个时代最美丽的建筑作品。有时候，玻璃的透明和钢材的轻盈可以产生一些奇妙的视觉效果，反映出现代人对民主的期望：公众领域和透明度。

建筑的另一大挑战是克服重力作用，而钢结构提供了完美的结构解决方案。[10]

我在设计钢结构建筑时，曾试图使用尽可能少的建材。当我还是一个年轻的建筑学学生时，我崇拜密斯·凡·德罗的建筑和建筑细节。细节在那个时代是非常重要的。

后来我发现建筑不仅仅是细节问题，更是一套综合的设计解决方案，通过建筑的各个方面改善使用者的周边环境。因此，这个大问题变得更加重要（图6-21）。

因为建筑业不能像汽车工业、航空制造业乃至船舶制造业那样达到一个统一的质量标准水平，我更多地关注于治愈性建筑。

我开始在我的设计作品中更多地使用钢材。这是一种现代材料，同时因为这一材料可以循环利用，它成为我设计火车站和地铁站最爱用的建筑材料（图6-22～图6-25）。

我最好的设计案例是赫尔辛基火车站站台雨棚，这是一个国际建筑竞赛项目。在一个和莫尼奥、福斯特以及其他著名建筑师的设计竞赛中胜出，对我来说是一次美好的经历和经验（图6-26、图6-27）。

图6-21 Hansasilta, Helsinki, Sakari Aartelo and Esa Piironen, 1984 汉萨西尔塔天桥购物中心，赫尔辛基，埃萨·皮罗宁，萨卡利·阿尔泰洛，1984

图6-22 Leppävaara Exchange Terminal, Esa Piironen, 2002 大赫尔辛基地区埃斯波市，勒帕瓦拉（Leppavara）换乘中心，埃萨·皮罗宁，2002

图6-23 Koivukylä Railway Station, Esa Piironen, 2004 科伊福克尔火车站，埃萨·皮罗宁，2004

I liked to use less steel material in building the roofing, but the structural engineers couldn't come to a better solution. That is always a problem with our education; we architects have to trust on engineers in building.

Le Corbusier praised the engineering skills of the industrial era in his book *Vers une architecture*. Ships, cars and aircraft served as the ideals, made of steel, in architecture. However, Le Corbusier's own architecture was largely based on concrete.

Buckminster Fuller favoured steel structures and was known to ask his colleagues, how much their buildings weighed. Steel allows for light–weight and transparent structures.

When building with steel, it is crucial to understand the fundamentals of the material's behaviour. Steel withstands tensile stress better than compressive stress. This will always be a point of departure for the design idiom when building with steel.

Steel rusts, unless it is coated, typically with paint. Fireproof coating may also be used, which helps steel structures avoid collapse in a fire.

Even stainless steel rusts in harsh conditions. For this purpose, acid–proof steel was developed, which withstands the elements for longer.

COR–TEN steel is a more recent innovation amongst steel materials. The steel has acquired a rusty surface which also protects the element from further atmospheric corrosion. The beautiful surface colour is reminiscent of deep copper.

More noble metals, such a copper, bronze and titanium (as used in Frank Gehry's Guggenheim Museum in Bilbao) also withstand natural elements well. These metals also age beautifully as they form a layer on the surface, the colour of which changes over time due to weathering. Copper, for example,

在建造屋顶时我喜欢少用一点钢材，但是结构工程师没法拿出一个更好的解决方案。这就是我们建筑教育中的一个问题，我们建筑师必须信任我们的结构工程师。

勒·柯布西耶在其《走向新建筑》中大力赞扬工业时代的工程技术。船舶、汽车和飞机是钢材制造的理想建筑。然而，柯布西耶自己的建筑大部分都是混凝土的。

巴克敏斯特·富勒喜欢钢结构建筑，最著名的就是他问同事，他们的建筑有多重。钢材使得轻盈且清晰的结构成为可能。

用钢材做建筑材料，关键是了解钢材性能的原理。钢材承受拉应力的能力低于承受压应力的能力。这一直都是寻求钢材建筑设计风格的出发点。

钢材易生锈，除非进行表面处理，一般是刷漆。防火外层常使用钢材，这有助于钢结构建筑在火灾中避免全面倒塌。即使是不锈钢在恶劣环境下也会生锈。因此，可以长时间防止生锈的耐酸钢被开发出来。

耐候钢是钢材的最新产品。这种钢材表面有一层保护锈层，可以保护钢材受到进一步的大气腐蚀。这种钢材的表面是一种漂亮的怀旧的深褐色。

其他贵金属，如铜、青铜、钛（弗兰克·盖里在西班牙毕尔巴鄂的古根海姆博物馆设计中使用了这种材料）也可以抗腐蚀。这些金属的表面层具有年代

图6-24　Mäntsälä Railway Station, Esa Piironen, 2006　曼塔萨拉火车站，埃萨·皮罗宁，2006

图6-25　Aalto University Metro Station, ALA+ESA, 2017　阿尔托大学地铁站，ALA+ESA, 2017

图6-26　Helsinki Railway Station roofing, Esa Piironen, 2001　赫尔辛基火车站雨棚，埃萨·皮罗宁，2001

图6-27　Helsinki Railway Station platform roofing, Esa Piironen, 2001　赫尔辛基火车站，埃萨·皮罗宁，2001

turns to a stunning green in a maritime climate while bronze blackens. This type of patina can also be achieved faster through artificial means.

We all encountered steel in our childhood in many forms. Perhaps the earliest and most memorable experience for many was sticking one's tongue on steel in freezing weather to see if it would stick. It might have been frightening, but all you had to do was to breathe warm air through your nose to release the tongue. The memory, unlike the tongue, stuck rests there for the rest of life.

Steel can feel warm or chilly, depending on the situation. In direct sunlight, steel can become scorching hot. In freezing weather, it feels equally cold.

In warm indoor spaces, hand rests, door handles, water taps and other similar items are often metal. Brass handles leave a metallic smell on the hands. Medical research has discovered that by using more noble metals in door handles and water taps (e.g. copper), the amount of bacterial build–up can be dramatically reduced in places such as hospitals and schools. This would have a positive effect on public health.

Steel is currently undergoing a phase of strong innovation work. The advantages of steel as a building material are the speed of building, its light weight, versatility, measurement accuracy and adaptability. From the ecological perspective, non–renewable building materials should be used as little as possible to save natural resources. Steel structures should be light–weight, slender with a gossamer thin appearance.

Recycling is a key factor in sustainable building. Steel can be recycled. Even the early examples, such as Paxton's Crystal Palace, was relocated with relative ease. Currently, nearly 100 per cent of all steel used has been re–melted.

One of the tasks of architecture is to defy gravity. Steel offers one of the best means of doing so.[11]

美感，由于气候变化，金属表面的颜色会随着时间的发展产生变化。例如，在海洋性气候环境下，铜的表面会变成漂亮的绿色而青铜会变黑。这种铜绿表面也可以通过人为的干预而在更短时间内获得。

在少年时期，我们都接触过不同形式的钢材。可能对许多人来说，最早的印象最深刻的经历是在冰冻天气中用舌头舔钢管，看看舌头会不会被黏住。这听起来可能有些可怕，但你只要用鼻子呼出热气来温暖融化你的舌头就可以了。和舌头不同，这些记忆会牢牢黏在你的生命中。

钢材的触感可以温暖也可以寒冷，取决于环境状况。在太阳直射下，钢材可以变得异常灼热。在寒冷的气候条件下，钢材就变得冰冷。

在温暖的室内空间，扶手、门把手、自来水龙头和其他一些相似的配件常常用金属。黄铜把手会在手上留下一股金属的味道。医学研究表明，在医院或学校这样的公共建筑中，门把手、水龙头等多用贵金属材料（如铜）可以大大减少细菌聚集的数量。这对公众健康有积极正面的作用。

钢铁业目前正处于强有力的创新阶段。钢材作为建筑材料的优点是建设速度快、重量轻、通用性强、尺度精确和适应性广。从生态学的观点看，不可再生的建筑材料应尽可能地少，以保护自然资源。钢结构应该具有轻盈、纤细而轻薄的外观。

再利用是可持续建筑的一个关键因素。钢材可以再利用。哪怕是早期的如水晶宫这样的钢结构，也可以相对容易地易地重建。现在，几乎100%的钢材都可以回收再利用。

建筑的主要任务之一是克服重力作用。钢结构提供了最好的解决办法之一。[11]

Stainless steel

Stainless steel is an alloy that contains less than 1.2% carbon and more than 12% chromium. Thanks to the chromium content, the surface of stainless steel undergoes passivation, forming an inert film of chromium oxide on the surface. This layer prevents further corrosion.

Ordinary stainless steel does, in fact, rust in certain conditions but acid–proof stainless steel has even greater resistance to corrosion in demanding weather conditions.

COR-TEN

Weathering steel (COR–TEN steel) is a low alloy steel which contains copper and chromium and forms an oxide layer on its surface under the influence of the weather, protecting it from corrosion.

Aluminum

Aluminium, an element abundant in the Earth's crust, is three times lighter than steel and yet its strength is still close to that of steel.

An even more durable duralumin has been created from aluminium. This alloy is used, for example, in aircraft.

When electronic microscopes enabled scientists to see inside metal crystals, it became possible to manipulate metals on the micro level. This innovation led to the development of super alloys with the aid of gamma particles. Super alloys have applications in, for example, jet engines. In this case, the metals are monocrystalline.

Aluminum is the most common metal in the world, more than twice as common as iron. Aluminum comes combined with a wide range of minerals, but it is economically feasible only when extracted from bauxite ore.

不锈钢

不锈钢是一种含碳量低于 1.2%、含铬量超过 12% 的合金。由于其铬成分，不锈钢表面经过钝化，形成了氧化铬惰性膜表层。这个表层可以防止进一步的腐蚀。

事实上，普通的不锈钢在一定的气候条件下会被腐蚀生锈，但耐酸不锈钢则有更强的抗腐蚀性。

耐候钢

耐候钢是一种含有铜和铬的低合金钢，在气候影响下形成的氧化惰性膜表层，保护其不受空气的进一步腐蚀。

铝

铝是地壳里含量丰富的一种元素，重量是钢的三分之一，却和钢的强度接近。

铝能制成更耐久的铝合金，例如，这种合金在飞机制造业中常被使用。

电子显微镜让科学家能够看到金属的内部晶体结构，这使得人们可以在微米层面处理金属材料。借助伽马（γ）粒子，这一革新带来了超级合金材料。例如，超级合金已经运用在喷气发动机上。在超级合金中，金属是单晶体。

铝是世界上最普通的金属，比铁还要普通。铝可以结合在很多种矿物中，但是只有从铝矾土矿中提取铝，从经济上讲是可行的。

Large areas of Amazonian rain forest (180,000 hectares/year) are cut down for this purpose. Bauxite is strip–mined from a depth of two meters.

While rain forests are being replanted, it is not fully known how this process will affect the earth's climate in the long term. Bauxite also provides raw materials for concrete manufacturing.

Red mud is the waste product generated in the industrial production of aluminum. No further uses have been developed for this highly alkaline waste product. Aluminum manufacturing consumes ten times as much energy as steel manufacturing.

Aluminum can be recycled and its value is multiplied five–fold at each stage. Aluminum products can be seen everywhere, from deodorants to sun creams. According to latest studies, it may, however, have dangerous side–effects, and increase the risk of, for example, Alzheimer's disease and breast cancer.[12]

Titanium

Titanium is rarely used as a building material. Its main use is in the aviation industry.

Frank Gehry managed to purchase titanium for a reasonable price when the Guggenheim Museum in Bilbao was built. The end result is a success both in terms of architectural substance and construction technology. The City of Bilbao reached the consciousness of the whole world and it continues to attract a large number of tourists every year.

Copper

Copper is a soft, malleable, and ductile metal with very high thermal and electrical conductivity. A freshly exposed surface of a pure copper has a reddish–orange color. Copper is used as a conductor of heat and electricity, as a building material and as a constituent of various metal alloys.

亚马逊雨林的大部分区域（180 000公顷/年）正因为铝矾土矿而被砍伐。铝土矿一般在两米深以下，呈带状进行开采。

现在雨林正在重新种植恢复中，但是从长期来看，这样的砍伐开采与恢复的整个进程对于地球气候变化的影响，人们还不是特别清楚。

铝矾土矿也为混凝土制造业提供原料。人们对铝矾土矿中的高碱性废弃物还没有开发出进一步的利用手段。

铝制造业会消耗钢制造业十倍的能源。铝可以回收利用，每再加工一次，其价值就是前一次的五倍。铝制品到处可见，从除臭剂到防晒霜。然而，根据最新的研究，铝有危险的副作用，例如提高阿尔茨海默病和乳腺癌的患病风险。[12]

钛

钛已经被用作建筑材料。钛主要是运用在航空业上。

弗兰克·盖里在西班牙毕尔巴鄂建造古根海姆博物馆时设法用非常合适的价格购买了钛材料。这个设计最终在建筑材料和结构技术上都非常成功。毕尔巴鄂这个城市因此得到了全世界的关注并且每年持续吸引着大批的游客。

铜

铜是一种柔软的、有延展性的并具有很高的导热性和导电性的韧性金属。纯铜新暴露的外表面呈橙红色。铜可以作为电和热的导体，可以做建筑材料以及各种合金的原材料。

Copper is one of the few metals that occur in nature in directly usable metallic form as opposed to needing extraction from an ore. This led to very early human use, from c. 8000 BC. It was the first metal to be smelted from its ore, c. 5000 BC, the first metal to be cast into a shape in a mold, c. 4000 BC and the first metal to be purposefully alloyed with another metal , tin to create bronze, c. 3500 BC.

In the Roman era, copper was principally mined in Cyprus, where the origin of the name comes.

Copper used in buildings, usually for roofing, oxidizes to form a green verdigris (or patina). Copper is sometimes used in decorative art, both in its elemental metal form and in compounds as pigments.

Copper has been used since ancient times as a durable, corrosion resistant, and weatherproof architectural material. Roofs, flashings, rain gutters, downspouts, domes, spires, vaults and doors have been made from copper for hundreds or thousands of years.

Copper's architectural use has been expanded in modern times to include interior and exterior wall cladding, building expansion joints and decorative indoor products such as attractive handrails, bathroom fixtures, and counter tops.

Some of copper's other important benefits as an architectural material include low thermal movement, light weight, lightning protection, and recyclability.

The metal's distinctive natural green patina has long been coveted by architects and designers. The final patina is a particularly durable layer that is highly resistant to atmospheric corrosion, thereby protecting the underlying metal against further weathering. Architectural copper and its alloys can also be 'finished' to embark a particular look, feel, and color. Finishes include mechanical surface treatments, chemical coloring, and coatings.

不同于其他金属需要从矿石中提取，铜是少数几种以金属形式存在于自然中可以直接利用的。这使得很早以前约公元前 8000 年，人类就开始使用铜。公元前 5000 年，人类第一次从矿石中冶炼金属；公元前 4000 年，人类第一次用模具将金属铸造成形；公元前 3500 年，人类第一次有目地将两种金属合成合金，锡合铜合成青铜。

在古罗马时期，铜主要在塞浦路斯开采，这也是铜这一名字的最初来源。

铜用在建筑上，主要是屋顶面，被氧化后形成绿色铜锈（铜绿）。铜有时被用在装饰艺术上，以其基本金属形式或者颜料的合成物这两种使用方式。

铜自古以来一直被作为一种耐用、耐腐蚀、耐候的建筑材料。用铜做的屋顶、遮雨板、雨水槽、水落管、穹顶、尖顶、拱顶和门等已有数百上千年的历史。现在铜在建筑上的用途扩展到了包括内外墙挂板、建筑收缩缝和诸如漂亮的把手、卫生间用具和台面板等室内装饰产品。

铜作为建筑材料的其他主要优点有较低的热运动、质轻、防雷和可循环再利用。

一直以来，建筑师和设计师们都对金属铜所产生的独特的铜绿梦寐以求。这层铜绿是一个耐久的外层，能防止大气腐蚀，从而保护铜内部不受进一步的侵蚀。建筑铜材和铜合金表面可以通过加工呈现出特别的外观、感觉和颜色。加工包括机械化的表面处理、化学染色和涂层。

Bronze

Bronze is an alloy consisting of copper, commonly with about 12% tin and often with the addition of other metals (such as aluminum, manganese, nickel or zinc) and sometimes non–metals or metalloids such as arsenic, phosphorus or silicon. These additions produce a range of alloys that may be harder than copper alone, or have other useful properties, such as stiffness, ductility, or machinability.

The archeological period where bronze widespread use is known as the Bronze Age. In the ancient Near East this began with the rise of Sumer in the 4th millennium BC, with India and China starting to use bronze around the same time; everywhere it gradually spread across regions.

Bronzes are typically very ductile alloys. By way of comparison, most bronzes are considerably less brittle than cast iron. Typically bronze only oxidizes superficially; once a copper oxide (eventually becoming copper carbonate) layer is formed, the underlying metal is protected from further corrosion.

Copper–based alloys have lower melting points than steel or iron, and are more readily produced from their constituent metals. They are generally about 10 percent denser than steel, although alloys using aluminum or silicon may be slightly less dense. Bronzes have lower hardness, strength and elastic modulus– bronze springs.

Bronze is widely used for casting sculptures. In modern architecture bronze is used with copper in exterior claddings and detail profiles worldwide. When bronze oxidizes, it will get a dark, black appearance. That is also possible to do it chemically for architectural reasons.

A building is time in the form of material.

Frank O. Gehry

青铜

青铜是一种铜合金，一般含有 12% 的锡，常加入其他金属（如铝、锰、镍、锌），或者有时加入非金属如砷、磷、硅等。这些不同的其他元素的加入形成了一系列铜合金，有些比铜坚硬，有些具有其他有用的特性，如高硬度、延展性或可加工性。

考古学中，青铜广泛使用的时期被称为青铜器时代。在古代近东青铜器时代始于公元前 4000 年苏美尔人的崛起，印度和中国大约在同一时期开始使用青铜；之后逐渐蔓延到世界各地。

青铜是典型的耐久合金。相比较而言，铸铁比大部分青铜要易碎。通常青铜只有表面被氧化，一旦氧化铜（最终形成碳酸铜）表面形成，金属内部就会被保护不受进一步的腐蚀。

铜合金的熔点一般比钢和铁低，且更易生产。它的密度一般比钢约高 10%，尽管使用铝或硅的合金密度可能稍低一些。

青铜广泛用于铸造雕塑。现代建筑中，青铜和铜被广泛用于外墙表皮和轮廓细部。当青铜氧化后颜色会变深。在建筑中使用时，也可以通过化学方式达到这样的效果。

建筑是其所在时代材料的形式物化。

——弗兰克·盖里

Chapter 7 Transparent as glass

Transparency enables the illusion
in architecture, which is part of the
essence of architecture.

Esa Piironen

Glass is an amorphous solid material formed from melted silicates. Glass is fragile yet hard and usually transparent.

The transparency of glass is what has always fascinated people. To see through a solid material was a new experience. The ability to cover or shutter this transparency has enabled a new type of architectural thinking.

Today, glass can be manufactured to be virtually unbreakable. By adding layers of glass, a pane will be able to withstand even a bazooka.

Reflection has always intrigued architects. Glass as a slightly reflective material creates interesting views; you can see through glass but also the reflection of the opposite view.

Fully opaque, reflective façades are uninviting, although they are widely used in skyscrapers' curtain walls.

第七章　通透之玻璃

透明能使人产生幻觉，这是建筑
本质的一部分。

——埃萨·皮罗宁

玻璃是由熔融的硅酸盐组成的非晶体固体材料。
玻璃坚硬易碎，通常呈透明状。

透明性一直是玻璃让人们感兴趣的原因。透过一
个固体材料看世界是一种全新的体验。如何遮盖
这一透明空间，使得形成新的建筑思维模式成为
可能。

今天，可以生产出坚固不易破裂的玻璃。多层玻
璃窗甚至可以承受住反坦克火箭筒的强大火力。

镜像常常令建筑师着迷。玻璃作为反光材料能创
造出富有趣味的景象；你可以透过玻璃观察，也
可以从玻璃上看到反射的景象（图7-1、图7-2）。

不透明的反射面材毫无魅力，虽然现在的摩天大楼
幕墙面上广泛使用这样的材料。

图7-1　Maison de Verre, Paris,
Pierre Chareau and Bernard
Bijvoet, 1931　玻璃之家，巴黎，
皮埃尔·沙罗和贝尔纳·比伊伏特，
1931

图7-2　Private Swimming Pool,
Turku, Esa Piironen and Mikko
Pulkkinen, 1968　私人泳池，图尔
库，埃萨·皮罗宁和米科·普尔基
宁，1968

Looking through two, three or even more several glass walls, placed at an angle in relation to each other may create strange visual sensations. Birds cannot comprehend this and become confused, often flying into glass, with fatal consequences.

Glass is an ancient innovation. In Egypt, alkali–lime glass was used to make beads as early as in 4000 BC. Glass was for long a very rarely used building material. Primitive dwellings had no windows. By 300 AD, when glass–making techniques had improved, Roman houses introduced glass window panes. The first glass window in England was made in 674 AD.

Medieval churches show the wealth that churches at that time had, to be able to afford stained glass windows to teach the congregation about biblical stories and events.

Stained glass art was also used for spiritual improvement.

A significant proportion of today's glass raw material is recycled.

Natural light is crucial for people's wellbeing. Even in the north, where daylight is scarce in the winter, all available natural light needs to be captured. Transparency enables an illusion, an integral part of architecture.

Architects had long dreamt of designing buildings made exclusively of glass. Mies van der Rohe gradually developed his dream of a fully glazed skyscraper around the 1920s. Since then, tall buildings have been built more or less from glass.

透过两层、三层甚至数层相互呈角度设置的玻璃幕墙，人们可以得到神奇的视觉感受。鸟类一般会被迷惑到无法分辨这类空间的真伪，一旦飞入这样的玻璃空间，会产生危险致命的后果。

玻璃是一种古老的发明。早在公元前4000年的埃及，碱石灰玻璃就被用来做成项链珠子。玻璃长久以来都是很不常用的建筑材料。原始民居是没有窗子的。直到公元300年，玻璃制造技术得到提高，古罗马人的房屋开始引入玻璃窗。在英格兰，最早的玻璃窗出现在公元674年。

中世纪的教堂，以有能力安装向民众传播圣经故事和事件的彩绘玻璃来显示这个教堂的财力。

彩绘玻璃艺术也可以带来心灵的升华。

今天玻璃原料的一个重要特性是可循环使用。

自然光对于人类健康至关重要。尤其在北方，冬日的自然光照时间短，需要抓住所有可以利用的自然光。透明性成就了建筑的多维空间，是健康建筑必不可少的组成部分。

设计仅用玻璃建造的建筑是建筑师们一直以来的梦想（图7-3、图7-4）。约20世纪20年代，密斯·凡德罗终于实现了一个全玻璃幕墙摩天楼的梦想。从那时起，高层建筑或多或少会用玻璃幕墙（图7-5）。

图7-3 Dancing Hall, Yläne, Ola Laiho, 1968 舞蹈厅，伊兰，奥拉莱霍，1968

图7-4 Willis Faber, Dumas_Norman Foster, 1975 威利斯·费伯和杜马总部大楼 诺曼·福斯特，1975

图7-5 Crown Hall, Ludwig Mies van der Rohe, 1956 克朗楼，密斯·凡·德罗，1956

Farnsworth House, the iconic glass box by Mies van der Rohe from 1950 has, since its completion, served as a model for many houses around the world.

One of them is the Glass House in New Canaan designed by Philip Johnson for himself. According to an anecdote, Frank Lloyd Wright once paid a visit to Philip Johnson's Glass House. When inside, he did not take his hat off. When Johnson asked him why he left his hat on, Wright replied: I don't know whether I'm in or out.

Modern technology has come up with self–cleaning glass, which is based on nanotechnology.

Electrically heated glass is practical in spaces where the chill radiating through large window panes would make sitting nearby uncomfortable. Thin, transparent metal films sandwiched between two panes conducts electricity providing heating.

A similar method is used for glass that can be darkened with the press of a button.

Glass may be tempered to better withstand impact.

Safety glass is made by adding a plastic film in between glass panes. If the glass sustains an impact, the glass may break but the plastic film keeps the splinters together. Tempered glass breaks into tiny rubble.

Glass is a bewitching material that is capable of transmitting, reflecting and conducting light. Glass looks black in daylight, but in nocturnal darkness it disappears completely. Its ethereal quality stimulates the imagination and its mineral sharpness alerts us to the vulnerability of our own bodies.

范斯沃斯住宅，密斯·凡·德罗设计的教科书式的玻璃盒子建筑，自从1950年竣工后就成为世界各地住宅设计的模板。

其中一个是菲利普·约翰逊在新卡纳为自己设计的天堂玻璃屋（图7-6）。有一则轶事说，弗兰克·劳埃德·赖特有一次去参观菲利普的玻璃屋，他进门后没有脱帽。当约翰逊问他为什么还戴着帽子时，赖特说：我不知道我有没有进屋。

图7-6　Glass House, Philip Johnson, 1949　天堂玻璃屋，菲利浦·约翰逊，1949

现代技术的发展带来了基于纳米技术的带自净功能的玻璃。

图7-7　Crystal Cathedral, Garden Grove, Philip Johnson, 1980　加州水晶大教堂，菲利普·约翰逊，1980

由于冷空气通过大面积玻璃窗会使坐在窗边的人很不舒服，使用电热玻璃是非常实用的。薄而透明的导电金属薄膜夹在两层玻璃间通电供热。

相似的技术可以通过使用一个按钮来使玻璃变暗。

钢化玻璃具有更高的抗冲击强度（图7-7～图7-9）。

图7-8　Reina Sofia, Ian Ritchie, 1990　雷纳索菲亚现代艺术博物馆玻璃塔（电梯间），西班牙，伊恩·里奇，1990

安全玻璃是在两层玻璃之间夹一层塑料薄膜。如果玻璃受到冲击，玻璃可能破碎但塑料薄膜能让碎玻璃保持整块。钢化玻璃则会碎成不带尖角的卵石状小颗粒。

玻璃是一种能够传输光、反射光和传导光的迷人材料。玻璃在白天是黑的，在黑夜中则完全消失不见。玻璃缥缈的特性可激发我们的想象力，其矿物质的锐利则警示着我们自身的脆弱（图7-10）。

图7-9　Fondation Cartier, Paris, Jean Nouvel, 1994　卡地亚基金会，巴黎，让·努韦尔，1994

In building, glass has only been widely used during the twentieth century. The key property that glass has is its transparency, which has symbolic meaning, too. Glass also has other, unpredictable properties, so that the architectural end result may turn out to be quite different from the architect's intentions. Considering dark office building from the aesthetic point of view may be overshadowed by images of glass caskets and glass coffins.

Some of the best–known exponents of twentieth–century glass architecture, Mies van der Rohe, Bruno Taut and Paul Scheerbart, for example, were all inspired by the great greenhouses of the nineteenth century. The technical approaches evolved for these glasshouses, that still continue to be evolved further even today, are a fine example of the lengths of technological development called for by the use of glass. Solutions based on a controlled indoor climate began to be applied from the nineteenth century onwards in other building types, mainly those used for temporary occupation, such as exhibition buildings. Difficulties emerged, however, when glass envelopes began to be used in premises intended for long–term occupation. Vast amounts of energy are consumed in resolving problems that would not exist if large expanses of glass surface were not used. The overall energy consumption of individual buildings should, after all, be examined from a global perspective.

The glass manufacturing process is a complex one. Its treatment demands great technical virtuosity; it cannot be touched during the manufacturing stage because of the high temperatures involved. Because glass is never 100% pure, it transmits light in different ways; it can let rays of light through or reflect them back.

Glass has complex associations, too. Glass is ethereal in its clarity and appeals to the imagination, while at the same time, it is deadly sharp and alerts the senses. Fairy tales about glass slippers and crystal coffins tell us something about the immaterial quality of glass at the same time as they associate it with rigor mortis, the stiffness of death. Glass can cause time to stand still.

在建筑上，玻璃只是在20世纪才被广泛地应用。玻璃的关键特性是其透明性，这也具有一定的象征性意义（图7-11）。玻璃还有其他一些未知特性，这导致建筑的最终效果会和建筑师的意图完全不同。从美学观点来看，深色的办公楼外立面较玻璃幕墙办公楼而言显得黯然失色。

图7-10　Leipzig Fair, van Gerkan, Gerkan, Marg and Ritchie　莱比锡贸易博览中心　玛格和里奇

20世纪玻璃盒子建筑的倡导者，例如建筑师密斯·凡·德罗、布鲁诺·陶特和保罗·希尔巴特都是受到了19世纪大温室建筑的启发。这些温室建筑所涉及的技术方法，直至今天仍然在不断发展，是玻璃应用技术发展时间线的参考范例。基于室内小气候控制的解决方案，从19世纪开始应用于其他建筑类型，主要用于临时空间如展览性建筑。然而，当玻璃材质外壳一旦运用到长期使用的建筑上时，困难出现了。如果不使用大量的玻璃外表面，就不会存在为了解决问题而需要消耗大量能源的问题。归根结底，单个建筑的能源消耗总量应该有一个全球视野下的考量。

图7-11　Louvre Pyramid, Paris, I.M.Pei, 1989　卢浮宫玻璃金字塔，巴黎，贝聿铭，1989

玻璃的制造过程很复杂，它需要精湛的技术。由于高温烧制，在制造过程中不能触碰到。因为玻璃不能达到100%的纯度，它传输光有不同的方式：可以让光线穿透也可以反射光线（图7-12、图7-13）。

图7-12　Koivukyla Railway Station, Esa Piironen, 2004　芬兰科伊福克尔火车站，埃萨·皮罗宁，2004

玻璃也可以使人产生很多联想。玻璃纯净透明，引人遐想，同时玻璃致命的尖锐性又时刻警示着人体感官。童话故事里的水晶鞋和水晶棺告诉我们玻璃虚幻的非物质性同时也和僵硬的尸体、死亡联系在一起。玻璃可以使时间静止。

图7-13　Dancing House, Prag, Frank Gehry, 1996　舞蹈之家　普拉格，弗兰克·盖里，1996

In 1964 Colin Rowe and Robert Slutzky defined transparency through Cubist painting. Literal transparency (in a glass elevation) can be distinguished from phenomenal transparency (a two-dimensional structure that produces three-dimensional space in Cubism). Detlef Mertins points out in his article Transparency: Autonomy and Relationality, that the phenomenal transparency they favoured was to Rowe and Slutzky an intellectual construct derived from examining a glass surface from a distance and from a fixed viewpoint. According to Mertins, they had not understood the space-time dimension in movement defined by Sigfried Giedion. Jose Quetglas agrees with Mertins, pointing out that Mies van der Rohe favoured an isolated stage in his Barcelona Pavilion, which needed no one to set foot in the space. The two-dimensional space was like a stage-set that was designed to be looked at, not to be experienced.

In today's democratic society, transparency has such a powerful symbolic meaning that it is difficult to grasp the ambiguities and inconsistencies inherent in the use of glass. The conclusion to be drawn from this is that the concept of phenomenal transparency is linked with static state, or lack of movement, which does not help in understanding how literal transparency actually functions in spatial experience. Architecture is viewed as implicitly fragmented and abstract. Façade architecture is assessed in isolation from its surroundings or spatial arrangement. The precision of glass technology calls for painstaking structural design that erodes the energy available for considering the spaces themselves. This sort of static, technocratic concept of architecture has been predominant throughout the twentieth century, despite a certain admiration for external and internal spaces linked

1964年柯林·罗和罗伯特·斯拉茨基通过立体绘画定义了透明性。字面上的透明概念（一个玻璃立面）可以和现象上的透明概念（立体绘画中通过二维画面画出三维空间）区分出来。德特莱夫·梅斯在他的文章"透明性"中指出：自主性和关系性，罗和斯拉茨基所青睐的透明现象概念来自于从远处而且固定的视角来观察一个玻璃表面。根据梅斯的观点，他们并没有理解希格弗莱德·吉迪恩[1]所定义的运动中的多维时空。乔斯·奎格拉斯赞同梅斯的观点，指出密斯·凡·德罗在他的巴塞罗那德国馆中偏爱孤立的舞台形态，不需要人们进入到这个空间。这种二维空间如同舞台布置一般是设计来让人看的，而不是设计来让人进入体验的（图7-14）。

在今天的民主社会，透明性有了强大的象征意义，很难找到使用玻璃材料时的歧义象征（图7-15～图7-18）。从中我们可以得到这样的结论：透明现象概念与静态形象或者静止不动相关，无法帮助我们在空间体验时理解字面所说的透明概念的真正功能。建筑反馈到视觉感受会被破碎化和抽象化。建筑立面会在不考虑周边环境和空间布局环境的情况下被独立评估。玻璃工艺的高精度决定了艰辛的结构设计工作，损害了建筑空间本身的能量利用。建筑的这种静态、技术概念主导了整个20世纪，而忽视了本应被尊重的室内和室外空间之间存在的动态的联系。结果是出现了一系列的导致人

图7-14　Barcelona Pavilion, Ludwig Mies van der Rohe, 1929（1986）巴塞罗那德国馆，路德维希·密斯·凡·德罗，1929（1986）

图7-15　Vuosaari Metro Station, Esa Piironen, 1998　乌萨里地铁站，埃萨·皮罗宁 1998

图7-16　Helsinki Railway Station platform roofing, Esa Piironen, 2001　赫尔辛基火车站雨棚，埃萨·皮罗宁，2001

图7-17　Keilaniemi Metro Station, ALA+ ESA, 2017　凯拉尼米地铁站，ALA+ ESA, 2017

1　希格弗莱德·吉迪恩（Sigfried Giedion）：波希米亚裔瑞士历史学家及建筑评论家，他的作品《空间·时间·建筑》是一部极具影响力的关于现代建筑历史的标准作品。（译者注）

in dynamic relationships. The result is a succession of inhospitable, rigid glass boxes, which arouse feelings of dread.[13]

Glass became widely used in office buildings in the New York of the 1950s, first as opaque panels between strip windows and then as larger surfaces.

Eventually the current fashion evolved, in which the difference between the window and the wall in the varyingly reflective glazed curtain walls can not be discerned in daylight, but the truth is revealed when the lights are on at night. Then what we see is the most common or garden strip windows, with an opaque section below, starting approximately from desktop level. If, against all odds, the building truly has windows from floor to ceiling, the ideal of transparency of working will still remain mainly symbolic. In the evening, if anything, we only see lone cleaners at work.[14]

My goal is to make a building as
immaterial as possible.

Helmut Jahn

们产生恐惧怪异感受的生硬的、僵化的玻璃盒子建筑。[13]

图7-18　Pikku-Huopalahti Multipurpose Hall, Helsinki Esa Piironen, 1997　社区多功能厅，赫尔辛基，埃萨·皮罗宁，1997

20世纪50年代，玻璃在纽约的办公建筑中广泛使用，最先是在非透明墙面上镶嵌的玻璃窗然后是大面积的建筑表皮。

最终，目前的时尚潮流又发生了变化，在白天，反射的玻璃幕墙无法分辨出窗和墙，只有到了晚上亮起的灯光才会揭开真相。然后，我们看到的是最普通的条形窗，下方是大致桌面高度的不透明墙体部分。如果克服所有困难，这一建筑的窗子真的从地板到天花板，这个完美的透明效果也主要在其象征意义上。到晚上，如果能看到什么，那就只剩下清洁工在孤独地做清扫工作。[14]

我的目标是设计一个尽可能透明的（无形的）建筑。

——赫尔穆特·扬

Chapter 8　Mouldable as plastics

The critical factors of architecture are
nature, material and man.

Rogelio Salmona

The term "plastics" covers a range of synthetic or semi–synthetic organic condensation or polymerization products that can be molded or extruded into objects, films, or fibres.

Their name is derived from the fact that in their semi–liquid state they are malleable, or have the property of plasticity. Plastics vary immensely in heat tolerance, hardness and resiliency. Combined with this adaptability, the general uniformity of composition and lightness of plastics ensures their use in almost all industrial applications today.

High performance plastics such as EFTE (fluoropolymeren) have become an ideal building material due to its high abrasion resistance and chemical inertness. (Eden, Cornwall/Nicholas Grimshaw).

Plastics are a relatively new material in construction. Plastics is a generic terms for a wide range of synthetic materials developed in the 1900s.

Natural caoutchouc was refined into rubber by mixing it with other ingredients;

第八章　可塑之塑料

建筑的关键要素是：自然、材料和使用者。

——罗杰里奥·萨尔莫纳

"塑料"一词涵盖了一系列合成或半合成的可模塑或挤压成物品、薄膜或纤维的有机缩合或聚合产品。它们的名字来源于这样一个事实：在半液态时，它们是可塑的，或具有可塑性。塑料在耐热性、硬度和弹性方面千差外别。塑料的适应性、成分的普遍一致性和较轻的质量确保了它们现在几乎可以应用到所有工业用途上。

图 8-1　Eden, Cornwall, Nicholas Grimshaw, 2001　伊甸园项目，康沃尔，尼古拉斯·格里姆肖，2001

高品质塑料如 EFTE（乙烯 – 四氟乙烯聚合物）因为其高耐磨性和化学稳定性已成为一种理想的建筑材料（伊甸园项目，康沃尔/尼古拉斯·格里姆肖）（图 8-1）。

塑料是一种相对较新的建筑材料，是 20 世纪发展起来的一系列合成材料的总称（图 8-2）。

图 8-2　Olympic Stadium, Munich Frei Otto, 1972　慕尼黑奥林匹克体育场，弗雷·奥托[1]（德），1972

天然橡胶可以和其他成分一起精炼出橡胶。这种

1　弗雷·奥托（Frei Otto），德国建筑师，2015 年普利兹克建筑奖得主。

the modern varieties of this rubber are classified as plastics. Many plastics are based on organic materials, such as petrochemicals and their derivatives.

The first childhood associations with plastics are soap containers, celluloid, bakelite, guttapercha, which usually smelled bad, or at least strange. Then there came plastic washing basins, buckets and other household items. In the construction industry, plastics were introduced fairly late to be used as moisture barriers, instead of glass (acrylic and polycarbonate), heat insulation (styrofoam) and many other applications.

Plastic is easy to work and cast in different moulds. Plastics do not withstand heat, so their usability in construction is limited. Plastics were introduced in boat–building at the same time when glass–fibre reinforced plastic was developed. The new material was also used for building elements and even whole buildings.

A leading example of this in Finland was the FUTURO summer house, developed in the 1960s and resembling a UFO. Its manufacture came to a halt with the energy crisis, during which oil production was heavily controlled to prevent an excessive increase in the oil price.

Today, plastics are used in the manufacture of cruise liner cabins and bathroom pods. The real bad news today come from oceans. The waste plastics are polluting the oceans and create a real disaster situation for sea life.

现代橡胶种类也属于塑料。许多塑料都来自有机
材料，如石油化学产品及其衍生物。

图8-3 Futuro, Matti Suuronen, 1970
未来之家，马蒂·萨罗内，1970

童年和塑料的最早联系是肥皂盒、赛璐珞（制造文
具或玩具用的硝化纤维塑料）、胶木（热固性塑料，
广泛用作电绝缘材料、家具零件、日用品、工艺
品等）、古塔胶（牙医填料）等这些很难闻或者至
少看起来很奇怪的东西。然后是塑料袋、大洗澡
盆、大水桶和一些其他的家居用品。在建筑行业，
塑料较晚才被引入，可代替玻璃用作防潮层（丙烯
酸塑料[1]和聚碳酸酯板材[2]），可用作保温层（聚苯
乙烯泡沫塑料[3]），此外还有其他多项用途。

塑料易于加工，可在不同模具中铸塑。塑料不耐
热，因此在建筑中它们的用途有限。

芬兰的一个典型案例是"未来夏日之家"，该项目
开发于20世纪60年代，外形看起来像一个UFO
（图8-3）。这一项目在石油生产受到严格控制以
防止油价过度增长的能源危机时期停产了。今天，
塑料被用于制造游轮舱和浴室舱。

今天，真正的坏消息来自海洋。废弃的塑料制品
正在污染着海洋，并形成了海洋生命真正的灾难。

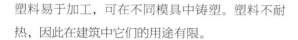

1 丙烯酸塑料：俗称的有机玻璃、亚克力板。（译者注）
2 聚碳酸酯板材：俗称PC板，具有良好的透光性、抗冲击性、耐紫外线辐射及其制品的尺寸稳定性
  和良好的成型加工性能，使其比建筑业传统使用的无机玻璃具有明显的技术性能优势。建筑上用
  作中空筋双壁板、暖房玻璃等。（译者注）
3 聚苯乙烯泡沫塑料：温度适应性强，在外墙保温中其占有率很高，是建筑专业中的一种B级材料。
  塑料被运用到造船业的同时，玻璃纤维强化塑料也在被开发利用。这一新型材料也被用作建筑元
  素，甚至被用于整栋建筑。（译者注）

Chapter 9   Foldable as fabric/textile

Architecture must understand, respect and
preserve equally people, the landscape and
materials.

Richard Leplastier

Fabrics have been used in building for almost as long as dwellings have been built. Tee–
pees in North America and yurts in Asia were among the first human shelters made from
fabrics (animal skin to be more precise).

The tent has been revived as a major construction technique with the development of tensile
architecture and synthetic fabrics.

Modern buildings can be made of flexible material such as fabric membranes, and supported
by a system of steel cables, rigid or internal, or by air pressure.

The circus tents are still being used all over the world. They are rapid to erect and pull down. The
atmosphere inside a circus tent can be restored for the rest of your life.

Only recently have fabrics been reintroduced in architecture, as new technologically
advanced textile materials have been developed. Tent constructions are among the newest
fields of expertise in engineering. Together with steel, fabric canopies can be used for
creating enormous span lengths.

第九章　可折之纤维/织物

建筑必须平等地理解、尊重、保护人、景观和材料。

——理查德·勒布拉斯蒂尔

织物被用于建筑领域差不多和住宅建造的历史一样长。北美印第安圆顶帐篷和亚洲的蒙古包是第一批用织物制成的人类住所（实际上应为动物的皮毛，比织物更为准确）。

随着张拉膜结构和合成纤维的发展，帐篷已作为一种重要的建筑技术得以复兴。

使用织物膜这一类柔性材料的现代建筑，可以通过刚性的、内部的或气压控制的钢索系统支撑。

马戏团帐篷仍在世界各地使用。这种帐篷易建易拆。马戏团帐篷里欢乐的氛围会永存你的记忆中，让你一生回味。

近几年，随着高技术纺织材料的发展，织物材料才被运用到建筑上。帐篷结构是工程专业技术的最新领域之一（图9-1）。结合钢材，织物可以用于建造大跨度建筑（图9-2～图9-5）。

图9-1　Huvila Tent, Helsinki, Roy Mänttäri, 1995　文化节大帐篷，赫尔辛基，罗伊·曼塔里，1995

图9-2　British Pavilion, Sevilla, Nicholas Grimshaw, 1992　英国馆，1992年西班牙塞维利亚世博会，尼古拉斯·格里姆肖，1992

Many airport concourses around the world are covered with a steel and fabric roofing. Textile can be used for creating impressive buildings, although their durability remains to be ascertained in the future. It is likely that textile–based structures need to be replaced from time to time, just like the wooden elements in Japanese temples. Increasing the life–cycle of textile materials in construction is still under development. Nanotechnology has the potential to present solutions on how to prolong the lifetime of materials.

世界上许多机场大厅都使用了钢架拉膜结构屋顶。织物可以创造出令人印象深刻的建筑,尽管其耐久性还有待将来进一步的试验确认。很可能织物结构的建筑需要不断地替换更新,如同日本寺庙中的木结构一般。延长织物建筑材料的寿命周期的研究开发工作仍在进行中。纳米技术在提供延长织物材料使用寿命的解决方案层面有很大潜力。

图9-3　Cricket Stadium, London, John Hopkins, 1997　板球运动场,伦敦,约翰·霍普金斯,1997

图 9-4　Savonlinna Opera Tent, Markku Erholtz and Heikki Paakkinen, 2000　萨冯林纳歌剧院顶棚,马尔库·埃尔霍尔茨里斯托·帕卡宁,2000

图9-5　Arch Defense, Paris, Johan Otto von Spreckelsen, 1989　拉·德方斯,巴黎,斯波莱克尔森(丹麦),1989

Postscript

In architecture, the spirit is more
important than the material.

Esa Piironen

There are millions of different materials in the world. Everything around us is made
of some material or other. There is a constant need for better materials, and there
are some rather exotic solutions being developed at the moment. New materials can
be smart, biodegradable, self–healing or, for example, electricity generating.

Combining different materials is being researched to achieve new functionalities,
such as surfaces that never get wet and always stay dry, or the ability to absorb or
repel bio molecules. Electricity and heat conductivity are also important qualities.
The goal is to create materials where the micro and nanostructure as such causes
the surface to repel water without the need for a separate coating. Here, the
problem lies in the mechanical durability of nanostructures.

Carbon, which a few decades ago was an old–fashioned fuel, has experienced
a revival, as new forms of carbon have been invented since the 1980s. Carbon
nanotubes, graphene, nano–diamonds and other innovations have already been
widely researched by the scientific community.

The difference between natural stone and synthetic materials has narrowed down
to a degree that it takes a specialist to tell them apart. Diamond–like carbon can
be used as a coating for a variety of objects. It can be used for reducing traction
between machine parts or inside plastic bottles as a gas barrier.

结语

建筑表达的精神涵义远
比建筑表现材料重要。

——埃萨·皮罗宁

世界上有成千上万种不同的材料。我们周围的每一件事物都是由某种或几种材料制成的。对于更好材料的需求一直存在，当下也发展出了一些相当奇特的材料解决方案。新材料可以是智能化的、能进行生物降解的、能自我恢复的或者自发电的……

人们正在研究不同材料的组合，以实现新的功能，例如不会被弄湿一直保持干燥的表层材料，或者能吸收或抵制生物分子的材料。导电性和导热性也是材料所需的重要品质。我们的目标是创造微米和纳米结构的新材料，使得表面防水不再需要加单独的涂层。在这里，问题在于纳米结构的机械耐久性上。

碳，几十年前是一种老式燃料，已经历了复兴。作为新形式的碳自 20 世纪 80 年代发明。碳纳米管、石墨烯、纳米钻石和其他的创新已经被科学界广泛研究。

天然石材和合成材料的差距已大大缩小，专家才能鉴定出它们的差别。类钻碳可以用作各种物体的涂层。它可以作为空气阻隔物减少机器零件之间或者玻璃瓶内的摩擦力。

Graphene has been applied to improve the durability of ceramic materials. The desirable qualities of ceramics–their hardness and heat resistance–come with a price, their brittleness: the slightest impact may break an entire ceramic mass into pieces. By layering aluminum oxide and graphene, a new material has been developed that bends but does not break.

Technological development has accelerated in the past few decades and keeps gaining even greater momentum.

Applications of 3D printing in construction are currently at a developmental stage, and they are expected to make rapid progress. For example, the Chinese WinSun has printed elements for a building several stories high from rapidly hardening concrete. The Finnish enterprise Fimatec has introduced the first 3D printed wall element.

The search is on for materials to replace reinforced concrete. Carbon neutral concrete is currently being investigated by several developers around the world. Rapidly hardening concrete suitable for 3D printed is being developed, as are fibre–based reinforcement materials to replace steel in achieving tensile strength and new fillers.

The development of construction robotics has progressed in leaps and bounds. American scientists have developed SAM, a bricklaying robot that lays bricks three times faster than a human bricklayer. Robots cannot replace humans but they can assist them. Robots can take over jobs that involve a lot of mechanical repetition. The bricklayer walks behind the robot and finishes the work, addresses visual details and cleans up excess mortar. Robots come into their own in building high walls.

Australians have developed a bricklaying robot that is able to lay the 15,000 bricks needed for a small house in two days.

石墨烯用于提高瓷砖的耐久性。坚硬耐热的高品质瓷砖价格很高，但瓷砖具有脆性：轻轻地一击会使整块瓷砖破碎成碎片。而通过加入铝氧化物和石墨烯层，人们开发出了一种能折弯而不会破碎的新型瓷砖。

新技术在过去的几十年间加速发展并且持续获得发展动力。

建筑 3D 打印技术正处在孕育发展阶段，有望取得快速的进步。例如，中国赢创（WinSun）用速干混凝土打印了多层建筑的各个单元。芬兰的 Fimatec 公司已经引入了首个 3D 打印墙体单元。

现在正在寻找钢筋混凝土的替代材料。世界各地许多研发者都在研究碳中性混凝土。适用于 3D 打印的速干混凝土正在开发，就像基于光纤的强化材料可代替钢材来实现拉伸强度和用作新型填料。

建造机器人技术迅速发展。美国科学家发明了 SAM，这种砌砖机器人的砌砖速度是熟练砌砖工人的三倍。机器人不能取代人类，但可以协助人类。机器人可以接管需要大量机械重复类型的工作。砌砖工人可以在机器人后面完成最后加工，完善装饰细节和清理多余的灰浆。在砌筑高墙时机器人可以行动自如。

澳大利亚人发明的砌砖机器人可以在两天内砌好一座小型房屋所需的 15000 块砖。

Alongside global warming, the most urgent problems affecting our planet are the disappearance of biodiversity and the accumulation of plastics in the oceans. Small plastic articles, originating from all kinds of plastic waste, have found their way into the oceans, where, when consumed by fish, can create and unforeseen problem for human beings.

All materials are ethically problematic. Architects must acknowledge their responsibility to the coming generations and aim to use environmentally sustainable materials when designing buildings.

New materials have always implied new architecture; nanotechnology is combining with biotechnology to create materials that are not only new, but that stand at the border of the animate and the inanimate, tending toward the living.

A corresponding architecture will surely follow. These materials must be seeded, grown and nurtured; so must the buildings that are formed by the same process.

Choosing the materials for a building is one of the key tasks of an architect. A building can incorporate a number of different materials, as the designer deems appropriate. Some materials work better in certain uses, while others are best employed somewhere else. The successful combination of materials requires visual and structural knowledge and insight from the architect. Sometimes, architects make the mistake of cramming too many different materials into one design, resulting in a building that lacks a cohesive idea. Other architects prefer to use very few materials, even striving towards creating single–material buildings. A case in point is the Japanese architect Tadao Ando, who is known for his heavy use of concrete. Concentrating on a single material in architecture lends it a certain gravitas.

The qualities of the materials and their application is heavily influenced by progress in information technology. Particularly since robots will be harnessed to an increasing degree in design and production engineering and new compound materials are introduced.

伴随全球变暖，影响地球最迫切的问题是生物多样性的消失和海洋中塑料物的堆积。源自各种塑料小制品的塑料废弃物，通过各种形式进入海洋，被海洋鱼类吞食，将会给人类带来无法预测的危害。

从伦理学上讲，目前的所有材料都是有问题的。建筑师必须认识到自己对于下一代的责任，在设计建筑时明确使用有利于环境的可持续材料。

新材料往往蕴含着新建筑的可能；纳米技术和生物技术的意义不仅是创造新材料，更是站在抉择人类生存和死亡的边界上。

相对应的建筑必定会出现。这些材料必须经过播种、成长和培育，因此建筑也必须通过同样的过程形成。

选择建筑材料是建筑师的关键工作之一。一栋建筑可以由建筑师认为合适的一系列不同的材料组成。某些材料在某种用途会更为合适，而其他材料则最好用于其他合适的地方。材料之间的成功衔接需要建筑师的视觉设计知识和建筑结构知识以及洞察力。有时候，建筑师犯的错误在于在一个设计中使用了过多种类的材料，导致这个建筑缺乏一个明确有力的主题。另一些建筑师则偏爱使用非常少数的几种建筑材料，甚至努力建造使用单一建材的建筑。相应的最佳案例是日本建筑师安藤忠雄，他以使用素混凝土著称。在建筑上专注于单一建材使得建筑有一定的庄严感。

材料的品质和应用深受信息技术进步的影响，特别是机器人在设计和生产工程中的不断增加和新型复合材料的使用。

It is important to already cast our eyes to the future and think about what a building will look like in thirty years from now.

That is when the real architectural value of a building is revealed.

Vitruvius' three tenets of architecture—firmitas, utilitas and venustas—are just as relevant today as then.

放眼未来，思考 30 年后的建筑会是什么样子非常重要。

那时，我们会看到建筑物真正的建筑学价值。

维特鲁威所说的建筑三原则——稳定性、实用性和美观性——在今天同样重要。

Appendix / 附录

Sources / 注释来源

注 [1] Piironen, Esa: Teräs julkisessa rakentamisessa, Rakennustieto, Helsinki, 1998

注 [2] Viljakainen, Mikko: wood calms you down and aids recovery, Puu 2/2014

注 [3] Siitonen, Tuomo: Mind and Matter, Arkkitehti 3/2016

注 [4] Mäkinen, Matti K: Arkkiteh+tuuri, Rakennustieto 2000

注 [5] Puu 3/2014

注 [6] Minke, Gernot: Building with Bamboo, 2016

注 [7] Mäkinen ibid

注 [8] Mäkinen ibid

注 [9] Pitkänen, Pekka: On Conrete Architecture, Arkkitehti 5/1967

注 [10] Piironen ibid

注 [11] Piironen ibid

注 [12] YLE/Teema 16.3.2015

注 [13] Blom, Kati: Illusions in the surface–transparency and doubt, Arkkitehti 2/2002

注 [14] Mäkinen ibid

Aforisms are from the book: *Thoughts On Architecture* edited by Esa Piironen, Fang Hai and Dongfang Tan, China Electric Power Press, 2018

本书中所有建筑师语录来自：埃萨·皮罗宁，方海，东方檀. 建筑哲思录 [M]. 北京：中国电力出版社，2018.

More Reading / 延展阅读

[1] Richard Weston. *Materials, Form and Architecture*, 2003.

[2] Sandu Cultural. Media ed. *Materials in Architecture*, 2012.

[3] Blaine Brownell. *Material Strategies: Innovative Applications in Architecture*, 2011.

[4] Axel Ritter. *Smart Materials in Architecture, Interior Architecture and Design*, 2006.

[5] Victoria Ballard Bell and Patrick Rand. *Materials for Design*, 2006.

[6] Rashida Ng and Sneha Patel. *Performative Materials in Architecture and Design*, 2013.

[7] Wendy Lesser. *You Say to Brick; The Life of Louis Kahn*, 2017.

[8] David Densie and Jacopo Gaspari. *Material Imagination in Architecture*, 2016.

[9] Bill Bryson. *At Home....*, 2010.

Images / 图片来源

All images are from the archives of Esa Piironen Architects.

Partly translations from Finnish to English by Verbum and Delingua, Helsinki.

所有图片均来自作者埃萨·皮罗宁本人。

从芬兰语译为英语的部分文字来自费尔堡与德林古（赫尔辛基）。

Index of Images / 图片索引

温暖如木 | Warm as wood

图 1-1　Old Chinese wood architecture　中国传统建筑木结构　039

图 1-2　Petäjävesi Church, Jaakko Leppänen, 1764　佩塔亚维西教堂，詹姆斯·列帕宁，1764　039

图 1-3　Shukagui Teahouse, Japan, 1659　虞姬茶室，日本，1659　039

图 1-4　Otaniemi chapel, Kaija and Heikki Siren, 1957　奥塔涅米礼拜堂，盖亚、海基西伦，1957　041

图 1-5　Yale Center for British Art, Louis Kahn, 1977　耶鲁英国艺术中心，路易斯·康，1977　041

图 1-6　Old Porvoo wooden houses　波尔沃历史小镇的木屋　043

图 1-7　Sea Ranch, Charles Mooreetal, 1965

　　　　海上牧场 / 摩尔，美国加州查尔斯·摩尔度假别墅（木结构），1965　043

图 1-8　Villa Mairea, Alvar Aalto, 1939　玛利亚别墅，阿尔瓦·阿尔托，1939　043

图 1-9　IIE New York, Alvar Aalto, 1965　纽约国际教育协会主会堂，阿尔瓦·阿尔托，1965　045

柔韧之竹 | Bendable as bamboo

图 2-1　Bamboo world　世界产竹区地图　061

图 2-2　Barajas Terminal, Madrid, Richard Rogers, 2004

　　　　西班牙马德里巴拉哈斯国际机场航站楼，理查德·罗杰斯，2004　061

图 2-3　Bamboo forest　竹海　061

图 2-4　Tampere Hall, Sakari Aartelo and Esa Piironen, 1990

　　　　坦佩雷音乐厅萨卡利·阿尔泰洛、埃萨·皮罗宁，1990　065

坚硬之石 | Hard as stone

图 3-1　Great pyramid Giza, Hemiunu 2560 BCE　吉萨大金字塔（胡夫金字塔），赫米翁努 公元前 2560　067

图 3-2　Hatsepsut Temple, Senenmut 1500 BCE　哈特谢普苏特神庙，森内马特 公元前 1500　067

图 3-3　Luxor　卢克索神庙　067

图 3-4　Machu Picchu　马丘比丘，秘鲁　069

图 3-5　Parthenon, Athens, Iktinos and Kallikrates, 438 BCE

　　　　帕特农神庙 / 万神庙，雅典，伊克提诺斯和卡拉克拉特，公元前 438　069

图 3-6　Yale Rare Book Library, Gordon Bunshaft, 1963　耶鲁大学图书馆，戈登·邦沙夫特，1963　069

图 3-7　Whitney Museum, NYC, Marcel Breuer, 1963　惠特尼博物馆，纽约，马塞尔·布劳耶，1963　069

图 3-8　Temppeliaukio Church, Helsinki Timo and Tuomo Suomalainen, 1969

　　　　岩石教堂（坦佩利奥基奥教堂），添姆和杜姆苏马连宁兄弟，芬兰赫尔辛基，坦佩雷，1969　069

图 3-9　Finlandia Hall, Helsinki, Alvar Aalto, 1974　芬兰宫 阿尔瓦·阿尔托，1974　071

图 3-10　Essen Opera House, Alvar Aalto, 1985　埃森歌剧院，阿尔瓦·阿尔托，德国北莱茵威斯特法伦州　071

图 3-11　American Center, Paris, Frank Gehry, 1994　美国中心，巴黎，弗兰克盖里，1994　071

图 3-12　AT&T NYC, Philip Johnson, 1984　纽约（AT&T）大厦，飞利浦·约翰逊，1984　071

方正之砖 | Solid as brick

图 4–1　Rautavaara Church, Sakari Aartelo and Esa Piironen, 1982
　　　　劳塔瓦拉教堂，萨卡利·阿尔泰洛，埃萨·皮罗宁，1982　　　　　　　　　　077

图 4–2　Robie House, Frank Lloyd Wright, 1910　罗宾之家，弗兰克·劳埃德·赖特，1910　　079

图 4–3　MIT Dormitory, Alvar Aalto, 1948　麻省理工学生宿舍，阿尔瓦·阿尔托，1948　　079

图 4–4　MIT Dormitory, Alvar Aalto, 1948　麻省理工学生宿舍，阿尔瓦·阿尔托，1948　　079

图 4–5　MIT Dormitory, Alvar Aalto, 1948　麻省理工学生宿舍，阿尔瓦·阿尔托，1948　　079

图 4–6　IIT Chapel, Mies Van der Rohe, 1951　伊利诺伊理工学院克朗小教堂，密斯·凡·德罗，1951　　081

图 4–7　Säynätsalo Town Hall, Alvar Aalto, 1952　珊纳特赛罗市政厅，阿尔瓦·阿尔托，1952　　081

图 4–8　Muuratsalo summer House, Alvar Aalto, 1952
　　　　穆拉萨罗岛实验住宅 / 夏日别墅，阿尔瓦·阿尔托，1952　　　　　　　　　　081

图 4–9　Yale Laboratory, Philip Johnson, 1963　耶鲁微生物学大楼，菲利普·约翰逊，1963　　081

图 4–10　TKK Main Building, Alvar Aalto, 1964　赫尔辛基理工学院主楼阿尔瓦·阿尔托，1964　　083

图 4–11　Fredensborg Housing area, Jorn Utzon, 1963　弗雷登斯堡小区，约翰·伍重，1963　　083

图 4–12　Sydney Opera House, Jorn Utzon, 1973　悉尼歌剧院，约翰·伍重，1973　　083

图 4–13　Rautavaara Church, Sakari Aartelo and Esa Piironen, 1982
　　　　劳塔瓦拉教堂，萨卡利·阿尔泰洛，埃萨·皮罗宁，1982　　　　　　　　　　083

图 4–14　Kauniairen Church, KristianGullichsen, 1983　考尼艾斯滕教堂，克里斯蒂安·古利克森，1983　　085

图 4–15　Poleeni, Pieksämäki, Kristian Gullichsen, 1990
　　　　波莱维，皮耶克赛迈基，克里斯蒂安·古利克森，1990　　　　　　　　　　085

图 4–16　Office Rautio Espoo, Sakari Aartelo and Esa Piironen, 1990
　　　　艾斯堡劳蒂奥办公楼，萨卡利·阿尔泰洛，埃萨·皮罗宁，1990　　　　　　　085

图 4–17　Kauhajoki School of Domestic Economics, Esa Piironen, 1992
　　　　考哈约基国内经济学院，埃萨·皮罗宁，1992　　　　　　　　　　　　　085

坚固之混凝土 | Reinforced as concrete

图 5–1　Pantheon, Rome, Apollodoros Damascus, 124 AD
　　　　万神庙，罗马，阿波罗多罗斯 大马士革 公元 124　　　　　　　　　　　　095

图 5–2　Colosseum, Rome, 80 AD　大角斗场，罗马，公元 8 世纪　　　　　　　　095

图 5–3　Casa Battlo, Barcelona, Antonio Gaudi, 1906　巴特罗公寓，巴塞罗那，安东尼·高迪，1906　　095

图 5–4　Villa Savoye, Le Corbusier, 1931　萨伏伊别墅，勒柯布西耶，1931　　095

图 5–5　Fallingwater, Frank Floyd Whrigt, 1939　流水别墅，弗兰克·劳埃德·赖特，1939　　095

图 5–6　Schröder Huis, Gerrit Rietveld, 1924　施罗德赫伊斯住宅，赫里特·里特费尔德，1924　　097

图 5–7　Paimio Sanatatorium, Alvar Aalto, 1929　派米奥疗养院，阿尔托，1929　　097

图 5–8　TWA Terminal, NYC, Eero Saarinen, 1962　纽约肯尼迪机场 TWA 候机厅，沙里宁，1962　　097

图 5–9　Guggenheim Museum, NYC, Frank Floyd Wright, 1959
　　　　纽约古根海姆博物馆，弗兰克·劳埃德·赖特，1959　　　　　　　　　　097

图 5–10　Unite d'habitation, Marseilles, Le Corbusier, 1952　马赛公寓，勒·柯布西耶，1952　　097

图 5–11　Chapel at Ronchamp, Le Corbusier, 1955　朗香教堂，勒·柯布西耶，1955　　099

图 5-12   Salk Institute, La Jolla, Louis Kahn, 1963   索尔克研究所，拉霍亚，路易斯·康，1963   099

图 5-13   Chapel of Holy Cross, Turku, Pekka Pitkänen, 1967

           圣十字教堂，土尔库，佩克卡·皮坎嫩，1967   099

图 5-14   Chapel of Holy Cross, Turku, Pekka Pitkänen, 1967

           圣十字教堂，图尔库，佩卡·皮特凯宁，1967   099

图 5-15   Private House, Kukkapuro, Paloheimo & Ollila, 1969

           私人住宅，库卡普罗，帕洛黑莫和奥利拉，1969   099

图 5-16   Harvard Visual Art Center, Le Corbusier, 1963   哈佛视觉艺术中心，勒·柯布西耶，1963   101

图 5-17   Yale Architectural School, New Haven, Paul Rudolph, 1963   耶鲁建筑学校，纽黑文市，路易斯·康，

           1963   101

图 5-18   Koshino House, Tadao Ando, 1981   小筱邸（住宅），安滕忠雄，1981   101

纤巧之钢 | Thin as steel

图 6-1   Tower Bridge, London, 1894   伦敦桥的钢结构基础塔楼，1894   113

图 6-2   Coalbrookedale bridge, 1779   科尔布鲁克代尔大桥，1779   113

图 6-3   GUM, Moscow, Alexander Pomerantsev, 1893   古姆百货，莫斯科，亚历山大·波梅兰采夫，1893   115

图 6-4   Crystal Palace, London, Joseph Paxton, 1851   水晶宫，伦敦，约瑟夫·帕克斯顿，1851   115

图 6-5   Eiffel Tower, Gustaf Eiffel, 1889   埃菲尔铁塔，古斯塔夫·埃菲尔，1889   117

图 6-6   Galeria Vittorio Emannuele Ⅱ, Milan, Giuseppe Mengini, 1867

           维托里奥·埃马努埃莱二世购物长廊，米兰，1867   117

图 6-7   Bradbury Building, LA, George Wyman, 1893   布拉德伯里大楼，加州洛杉矶，乔治·怀曼，1893 119

图 6-8   Chyrler Bldg NYC, William van Alen, 1931   纽约克莱斯勒大厦，威廉·凡·艾伦，1931   119

图 6-9   Empire State Building, NYC, Shreve, Lamb and Harmon, 1931

           帝国大厦，纽约，施里夫，兰博 & 哈蒙建筑公司，1931   119

图 6-10   Sears Tower, Chigaco, SOM, 1937   西尔斯大厦，芝加哥，史欧姆公司设计，1937   121

图 6-11   Heidi Weber Gallery, Zurich, Le Corbusier, 1963

           海蒂·韦伯博物馆，苏黎世，勒·柯布西耶，1963   121

图 6-12   Seagram NYC, Mies van der Rohe, 1958   纽约西格拉姆大厦，密斯·凡·德罗，1958   123

图 6-13   Farnsworth House, Plano, Ludwig Mies van der Rohe, 1951

           范斯沃斯住宅，普莱诺，路德维希·密斯·凡·德罗，1951   123

图 6-14   House #8, Santa Monica, Ray and Charles Eames, 1949

           八号住宅，圣莫妮卡，雷与查尔斯·伊姆斯，1949   123

图 6-15   Pompidou Center, Paris, Renzo Piano and Richard Rogers, 1977

           蓬皮杜中心，巴黎，伦佐·皮亚诺、理查德·罗杰斯，1977   125

图 6-16   Lloyd's, London, Richard Rogers, 1986   劳埃德大厦，理查德·罗杰斯，1986   125

图 6-17   HSBC, HongKong, Norman Foster, 1986   汇丰银行总部大楼，诺曼·福斯特，1986   125

图 6-18   Waterloo Terminal, London, Nicholas Grimshaw, 1993

           滑铁卢车站，伦敦，尼古拉斯·格里姆肖，1993   125

图 6-19   National Library, Paris, Dominique Perrault, 1995   国家图书馆，巴黎，多米尼克·佩罗，1995   127

图 6-20   Guggenheim Bilbao, Frank Gehry, 1997   古根海姆博物馆，毕尔巴鄂·弗兰克·盖里，1997   127

图 6-21　Hansasilta, Helsinki, Sakari Aartelo and Esa Piironen, 1984

汉萨西尔塔天桥购物中心，赫尔辛基，埃萨·皮罗宁，萨卡利·阿尔泰洛，1984　　　129

图 6-22　Leppävaara Exchance Terminal, Esa Piironen, 2002

大赫尔辛基地区埃斯波市，勒帕瓦拉（Leppavara）换乘中心，埃萨·皮罗宁，2002　　　129

图 6-23　Koivukylä Railway Station, Esa Piironen, 2004　科伊福克尔火车站，埃萨·皮罗宁，2004　129

图 6-24　Mäntsälä Railway Station, Esa Piironen, 2006　曼塔萨拉火车站，埃萨·皮罗宁，2006　　131

图 6-25　Aalto University Metro Station, ALA+ESA, 2017　阿尔托大学地铁站，ALA+ESA，2017　　131

图 6-26　Helsinki Railway Station roofing, Esa Piironen, 2001

赫尔辛基火车站雨棚，埃萨·皮罗宁，2001　　　131

图 6-27　Helsinki Railway Station platform roofing, Esa Piironen, 2001

赫尔辛基火车站，埃萨·皮罗宁，2001　　　133

通透之玻璃 | Transparent as glass

图 7-1　Maison de Verre, Paris, Pierre Chareau and Bernard Bijvoet, 1931

玻璃之家，巴黎，皮埃尔·沙罗和贝尔纳·比伊伏特，1931　　　145

图 7-2　Private Swimming Pool, Turku, Esa Piironen and Mikko Pulkkinen, 1968

私人泳池，图尔库，埃萨·皮罗宁和米科·普尔基宁，1968　　　145

图 7-3　Dancing Hall, Yläne, Ola Laiho, 1968　舞蹈厅，伊兰，奥拉莱霍，1968　　　147

图 7-4　Willis Faber, Dumas_Norman Foster, 1975　威利斯·费伯和杜马总部大楼 诺曼·福斯特，1975　147

图 7-5　Crown Hall, Ludwig Mies van der Rohe, 1956　克朗楼，密斯·凡·德罗，1956　　　147

图 7-6　Glass House, Philip Johnson, 1949　天堂玻璃屋，菲利浦·约翰逊，1949　　　149

图 7-7　Crystal Cathedral, Garden Grove, Philip Johnson, 1980　加州水晶大教堂，菲利普·约翰逊，1980　149

图 7-8　Reina Sofia, Ian Ritchie, 1990

雷纳索菲亚现代艺术博物馆玻璃塔（电梯间），西班牙，伊恩·里奇，1990　　　149

图 7-9　Fondation Cartier, Paris, Jean Nouvel, 1994　卡地亚基金会，巴黎，让·努韦尔，1994　　149

图 7-10　Leipzig Fair, van Gerkan, Gerkan, Marg and Ritchie　莱比锡贸易博览中心 玛格和里奇　151

图 7-11　Louvre Pyramid, Paris, I.M.Pei, 1989　卢浮宫玻璃金字塔，巴黎，贝聿铭，1989　　151

图 7-12　Koivukyla Railway Station, Esa Piironen, 2004　芬兰科伊福克尔火车站，埃萨·皮罗宁，2004　151

图 7-13　Dancing House, Prag, Frank Gehry, 1996　舞蹈之家 普拉格，弗兰克·盖里，1996　　151

图 7-14　Barcelona Pavilion, Ludwig Mies van der Rohe, 1929（1986）

巴塞罗那德国馆，路德维希·密斯·凡·德罗，1929（1986）　　　153

图 7-15　Vuosaari Metro Station, Esa Piironen, 1998　乌萨里地铁站，埃萨·皮罗宁，1998　　153

图 7-16　Helsinki Railway Station platform roofing, Esa Piironen, 2001

赫尔辛基火车站雨棚，埃萨·皮罗宁，2001　　　153

图 7-17　Keilaniemi Metro Station, ALA+ ESA, 2017　凯拉尼米地铁站，ALA+ ESA，2017　　153

图 7-18　Pikku–Huopalahti Multipurpose Hall, Helsinki Esa Piironen, 1997

社区多功能厅，赫尔辛基，埃萨·皮罗宁，1997　　　155

可塑之塑料 | Mouldable as plastics

图 8–1　Eden, Cornwall, Nicholas Grimshaw, 2001　伊甸园项目，康沃尔，尼古拉斯·格里姆肖，2001　157

图 8–2　Olympic Stadium, Munich Frei Otto, 1972　慕尼黑奥林匹克体育场，弗雷·奥托（德），1972　157

图 8–3　Futuro, Matti Suuronen, 1970　未来之家，马蒂·萨罗内，1970　159

可折之纤维 / 织物 | Foldable as fabric/textile

图 9–1　Huvila Tent, Helsinki, Roy Mänttäri, 1995　文化节大帐篷，赫尔辛基，罗伊·曼塔里，1995　161

图 9–2　British Pavilion, Sevilla, Nicholas Grimshaw, 1992

英国馆，1992 年西班牙塞维利亚世博会，尼古拉斯·格里姆肖，1992　161

图 9–3　Cricket Stadium, London, John Hopkins, 1997　板球运动场，伦敦，约翰·霍普金斯，1997　163

图 9–4　Savonlinna Opera Tent,Markku Erholtz and Heikki Paakkinen, 2000

萨冯林纳歌剧院顶棚，马尔库·埃尔霍尔茨里斯托·帕卡宁，2000　163

图 9–5　Arch Defense, Paris, Johan Otto von Spreckelsen, 1989

拉·德方斯，巴黎，斯波莱克尔森（丹麦），1989　163

Keilaniemi Metro Station, ALA+ ESA, 2017　凯拉尼米地铁站 ALA+ ESA，2017　180

译后记
张亚萍

翻译芬兰建筑师埃萨·皮罗宁《建筑与材料》这本著作，我内心是十分欢喜的，因为出于对建筑史和建筑设计的热爱与兴趣，而芬兰建筑正是我最感兴趣的一块。翻译芬兰当代代表建筑师的作品，无疑是一个绝佳的学习机会。

我清晰地记得翻译此书的那段日子正值暑夏，不时出现40℃的高温，而我却沉浸在那间有着一台年代久远、外机嗡嗡作响的空调的小书房中，查阅着各种相关资料，努力翻译。当看到一摞初步成形的译稿，烈日中透出的清凉，还有汗水背后泛出的阵阵喜悦与满足，无法言表。

20世纪初，芬兰建筑师阿尔瓦·阿尔托的出现，令芬兰建筑被国际建筑界认可。此后，芬兰的建筑设计在世界建筑设计界长期占有领先的地位，并产生出一大批优秀的建筑师。这些后起的建筑师继承传统、接纳吸收世界潮流、勤于理论研究，使得芬兰的建筑设计一直处于领先并且被广泛认可的地位，值得学习！埃萨·皮罗宁正是这些后起建筑师中极具代表性的一位。他是芬兰当代理性主义倾向的建筑师代表，他重视设计手法的逻辑性与新技术新材料的应用，同时他对芬兰传统的审美观念有较为深入的了解，对芬兰建筑的传统抱有高涨的热情，关心世界建筑的发展潮流并热衷于将之与芬兰本土文化相结合。整个翻译的过程，能感受到的是埃萨·皮罗宁先生务实、尊重自然、不断学习、勇于创新的设计精神。

芬兰的森林资源丰富，在芬兰传统建筑中从结构材料到装饰材料，木材都扮演着